International Conference on Design Engineering and Science 2016 (ICDES 2016)

MATEC Web of Conferences Volume 52 (2016)

Kuala Lumpur, Malaysia
27 - 29 February 2016

Editor:

Cees de Bont

ISBN: 978-1-5108-2271-9

Printed from e-media with permission by:

Curran Associates, Inc.
57 Morehouse Lane
Red Hook, NY 12571

Some format issues inherent in the e-media version may also appear in this print version.

Printed by Curran Associates, Inc. (2016)

For additional information, please contact EDP Sciences – Web of Conferences at the address below.

EDP Sciences – Web of Conferences
17, Avenue du Hoggar
Parc d'Activité de Courtabœuf
BP 112
F-91944 Les Ulis Cedex A
France

Phone: +33 (0) 1 69 18 75 75
Fax: +33 (0) 1 69 28 84 91

contact-edps@webofconferences.org

Additional copies of this publication are available from:

Curran Associates, Inc.
57 Morehouse Lane
Red Hook, NY 12571 USA
Phone: 845-758-0400
Fax: 845-758-2633
Email: curran@proceedings.com
Web: www.proceedings.com

TABLE OF CONTENTS

AIRCRAFT PERFORMANCE ANALYSIS IN CONCEPTUAL DESIGN PHASE BASED ON SYSTEM-OF-SYSTEMS SIMULATIONS 1
Tian Yifeng, Liu Hu, Huang Jun

COMBINED CYCLE FATIGUE INVESTIGATION BASED ON ENERGY PRINCIPLE 8
Kalynenko Mykyta

CIRCUIT MODELING AND EM SIMULATION VERIFICATION OF DGS BASED LOW-PASS FILTER EMPLOYING TRANSMISSION LINE MODEL ALONG WITH MICROSTRIP-SLOTLINE TRANSITIONS 16
Challal Mouloud, Dehmas Mokrane, Azrar Arab, Aksas Rabia, Trabelsi Mohamed

REQUIREMENTS ENGINEERING MODEL: ROLE BASED GOAL ORIENTED MODEL 22
Sandfreni, Surendro Ir. Kridanto

ANT FORAGING BEHAVIOR FOR JOB SHOP PROBLEM 30
Mahad Diyana Abdul, Ismail Zuhaimy

COMPARATIVE CALIBRATION OF CORROSION MEASUREMENTS USING K-NEAREST NEIGHBOUR BASED TECHNIQUES 36
Hamed Yaman, Shafie A'fza, Bt Mustaffa Zahiraniza, Binti Idris Naila Rusma

A STUDY ON A MULTI-OBJECTIVE OPTIMIZATION METHOD BASED ON NEURO-RESPONSE SURFACE METHOD (NRSM) 45
Lee Jae-Chul, Jeong Ji-Ho, Kharoufi Hicham, Shin Sung-Chul

DESIGN OPTIMIZATION OF ESD (EMERGENCY SHUTDOWN) SYSTEM FOR OFFSHORE PROCESS BASED ON RELIABILITY ANALYSIS 55
Bae Jeong-hoon, Shin Sung-chul, Park Byeong-cheol, Kim Soo-young

PRODUCT SYSTEM DESIGN – TO HOUSEHOLD MASSAGE DESIGN AS AN EXAMPLE 65
Wang Huabin, Lin Qichao

EFFECT OF LOCAL CONSTRAINT ON DRIVING TORQUE OF DRIVING MECHANISM FOR HALF-ROTATING WING 70
Wang Huixing, Wang Xiaoyi, Qiu Han, Chen Fuqiang, Qiu Zhizhen, Shan Jianhua

RESEARCH ON THE APPLICATION OF TACTILE MATERIALS TO DESIGN AND PRINTING OF BOOKS 76
Wang Yuanyuan

ADAPTIVE ROBOTIC SYSTEMS DESIGN IN UNIVERSITY OF APPLIED SCIENCES 82
Gunsing Jos, Gijselhart Fons, Hagemans Nyke, Jonkers Hans, Kivits Eric, Klijn Peter, Kapteijns Bart, Kroeske Diederich, Langen Hans, Oerlemans Bart, Oostindie Jan, van Stuijvenberg Joost

A TOOTH FLANK CROWNING METHOD BY APPLYING A NOVEL CROSSED ANGLE FUNCTION BETWEEN THE HOB CUTTER AND WORK GEAR IN THE GEAR HOBBING PROCESS 88
Wu Yu-Ren, Tran Van-The, Hsu Ruei-Hung

IN-VITRO DESIGN PROTOCOL: ARTIFICIAL SITUATION STRATEGY USES TO COMPREHEND DESIGNERS' THOUGHT 95
Anwar Rusmadiah, Abidin Shahriman Zainal, Hassan Oskar Hasdinor

THE APPLICATION OF CLOUD INNOVATION IN INDUSTRIAL DESIGN EDUCATION 104
Jiang Li-jun, Song Bo, Su Jing, Li Zhe-lin

MODERN LANDSCAPE REPRESENTATION OF HAKKA CULTURE 110
Huang Hui, Gao Ruofei

Author Index

PREFACE

2016 International Conference on Design Engineering and Science (ICDES 2016) was hosted at University of Malaya in Kuala Lumpur, Malaysia during February 27-29, 2016. ICDES 2016 was organized by Hong Kong Society of Mechanical Engineers (HKSME).

The conference provides a useful platform both for showing the latest research directions and for exchange of research results and ideas in Design Engineering and Science areas. The success of ICDES 2016 is reflected in the papers received, with participants coming from several countries, allowing a real multinational exchange of experiences and ideas.

ICDES 2016 includes a total of 16 selected high quality papers related to Design Engineering and Science fields. Last but not least, we would like to thank the conference chairs, organization staff, the international scientific committees and the reviewers for their excellent work in reviewing the papers as well as their invaluable input and effort. Thanks also goes to the University of Malaya and EDP Sciences for their various support to ICDES 2016.

Prof. Cees de Bont

Hong Kong Polytechnic University, Hong Kong

ICDES 2016 Committee

Conference Committee
Prof. Cees de Bont , Hong Kong Polytechnic University, Hong Kong (Chair)
Prof. Fred van Houten, University of Twente, Netherlands

Program Committee
Prof. Lucienne Blessing, University of Luxembourg, Germany

Technical Committee
Prof. Kai Cheng, Brunel University, UK
Prof. Belal Yousif, University of Southern Queensland, Austrilia
Prof. Eric Coatanea , Tampere University of Technology, Finland
Prof. Jan Detand, Ghent University, Belgium
Prof. Ismail Fidan,Tennessee Tech University, USA
Prof. John Gero, George Mason University, USA
Prof. Alex Duffy, University of Strathclyde, UK
Prof. Andrew Rae, University of the Highlands and Islands, UK
Prof. LAU Kin Tak Alan, Hong Kong Polytechnic University, Hong Kong
Prof. Jos Gunsing, Avans University of Applied Science, Netherlands
Prof. Gaetano Cascini, Politecnico di Milano,Italy
Prof. Zaidi bin Mohd Ripin, Universiti Sains Malaysia, Malaysia
Dr. Mojtaba Ahmadi, Carelton University, Canada
Prof. Tetsuo Tomiyama , Cranfield University, UK
Dr. Tai Kang, Nanyang Technology University, Singapore
Dr. FU Ming Wang, Hong Kong Polytechnic University, Hong Kong
Dr. Moon Seung Ki, Nanyang Technology University, Singapore
Dr. Jian Jun Zhu , Sr. CAE in vehicle dynamics and simulation at General Motors, USA
Dr. Edzrol Niza Bin Mohamad, University of Malaya, Malaysia
Dr. Mansoor Alghooneh, Immersion Corporation, Canada
Prof. Henri Christiaans, Ulsan National Institute of Science & Technology, South Korea
Prof. Gopinath Chintala, Alliance University, India
Prof. Yu-Ren Wu, National Central University, Taiwan
Dr. S.C. Johnson Lim, Universiti Tun Hussein Onn Malaysia, Malaysia

edp sciences

MATEC Web of Conferences

Statement of Peer review

In submitting conference proceedings to MATEC Web of Conferences, I certify to the Publisher that I adhere to the **Policy on Publishing Integrity** of the journal in order to safeguard good scientific practice in publishing.

1. All articles have been subjected to peer review administered by the proceedings editors.
2. Reviews have been conducted by expert referees, who have been requested to provide unbiased and constructive comments aimed, whenever possible, at improving the work.
3. Proceedings editors have taken all reasonable steps to ensure the quality of the materials they publish and their decision to accept or reject a paper for publication has been based only on the merits of the work and the relevance to the journal.

Title, date and place of the conference

2016 International Conference on Design Engineering and Science (ICDES 2016)

Feb. 27-29, 2016 Kuala Lumpur, Malaysia

Proceedings editor(s):

Prof. Cees De Bont

Date and editor's signature

- 8 MAR 2016

17, avenue du Hoggar - PA de Courtabœuf – BP 112 - 91944 Les Ulis Cedex A (France)
Tél. : 33 (0)1 69 18 75 75 - Fax : 33(0)1 69 07 45 17 – www.edpsciences.org

MATEC Web of Conferences 52, 01001 (2016)
DOI: 10.1051/matecconf/20165201001

Aircraft Performance Analysis in Conceptual Design Phase based on System-of-Systems Simulations

Yifeng Tian[1,a], Hu Liu[1] and Jun Huang[1]

[1]*School of Aeronautic Science and Engineering, Beihang University, Beijing, China*

Abstract. To obtain a competitive design in aircraft conceptual design phase, this paper propose an aircraft performance analysis method based on system-of-systems (SoS) simulations. The analysis process includes design space exploration and mission effectiveness analysis and the stochastic nature in SoS mission is captured based on a Monte Carlo method. Meanwhile as an example, the performance of UAVs which are applied in earthquake search and rescue SoS is analysed based on SoS simulations. The results show that the level of aircraft performance which leads to an optimum SoS effectiveness could be designed below the maximum performance. And it is necessary to apply SoS simulations into aircraft conceptual design phase to help designers optimize aircraft performance.

1 Introduction

In the past, the top-down decisions about aircraft performance parameters in conceptual design phase are usually made by chief engineers based on their experience, intuition and judgment [1]. And the value of original target design parameters are often set according to the existing aircraft or systems. To some degree, the design space exploration, parameter optimization and life-cycle cost are neglected, often leading to substantially unnecessary time and cost [2, 3]. Meanwhile, as the complexity in the fields of civil and military aviation is continuously increasing, aircraft usually carry out large system-of-systems (SoS) missions which make aircraft performance design more complicated due to the competition of different systems. So performance parameters should be analysed in depth.

The concept of SoS was first proposed in 1956 [4], but it was not until 1989 that SoS was introduced to engineering application [5]. It are usually made up of a multitude of independent but interactive systems. In recent years, this field has been growing significantly and SoS theory has been applied in many aeronautical engineering problems, such as air strike [6, 7], civilian transportation [8, 9] and so on. Different methods have also been applied to deal with these SoS problems, such as surrogate models [6, 7], mixed integer nonlinear programming [8], network theory [9], Agent [10], etc. Although there is still a lack of a common definition in the primary stage of SoS theory [10], the characteristics proposed by Maier [11] have been widely recognized, i.e., emergent behaviour, evolutionary development, operational independence of the elements, managerial independence of the elements and geographic distribution. The first two characteristics could also be found in systems because systems are established to acquire new capacity. However, the latter three characteristics are

[a] Corresponding author: tianyifeng_buaa@163.com

rather unique. Because component systems in different geographic positions have a high autonomy but also interact with each other, the uncertainty and discontinuities increase the difficulties in analysing and predicting the results of SoS mission. So when aircraft carry out a SoS mission, mission effectiveness is not only affected by aircraft performance but also by other component systems. Due to the competition of different systems in an SoS, the aircraft with a low level of performance parameters may also lead to a high mission effectiveness. Thus the desired aircraft is not the one with maximum performance but the one which leads to an optimal mission effectiveness.

Although in conceptual design phase there is no existing aircraft to conduct experiments to study the relationship between aircraft performance and mission effectiveness, modelling and simulation in a computer provides a feasible way to creating the mapping from measure of performance to measure of effectiveness. Through establishing SoS model and operating simulation program repeatedly based on Monte Carlo simulations, large amount of data in the uncertainty environment could be obtained. These data can provide designers two aspects of assessment: exploring design space according to mission effectiveness and analysing mission effectiveness of different designs.

In order to illuminate the issue, an example about unmanned aerial vehicle (UAV) search and rescue SoS in earthquake relief is discussed in the following sections. The SoS model is built according to earthquake relief in the reality. Based on SoS simulations, the UAV performance is analysed from the perspective of both design space exploration and mission effectiveness analysis.

2 Modelling process of UAV search and rescue SoS

2.1 Operation logic of mission simulation

When establishing an SoS model, the operation logic of each component system and the interactive logic of different component systems are very important. In fact, the example UAV search and rescue SoS in earthquake relief is very complex. However, the purpose of this paper is mainly to illuminate the important role of SoS simulation in aircraft conceptual design phase, some reasonable simplifications are set to reduce the difficulty in modelling process. When establishing the models of component systems, UAVs, survivors, temporary base and helicopters are selected, regardless of ground rescue because the terrain is set as mountainous area and the mission time is limited. Meanwhile the weather system is not taken into account. However, if necessary, a more comprehensive SoS model could be built for mission simulations. Considering the autonomy of component systems, systems are modelled as Agent.

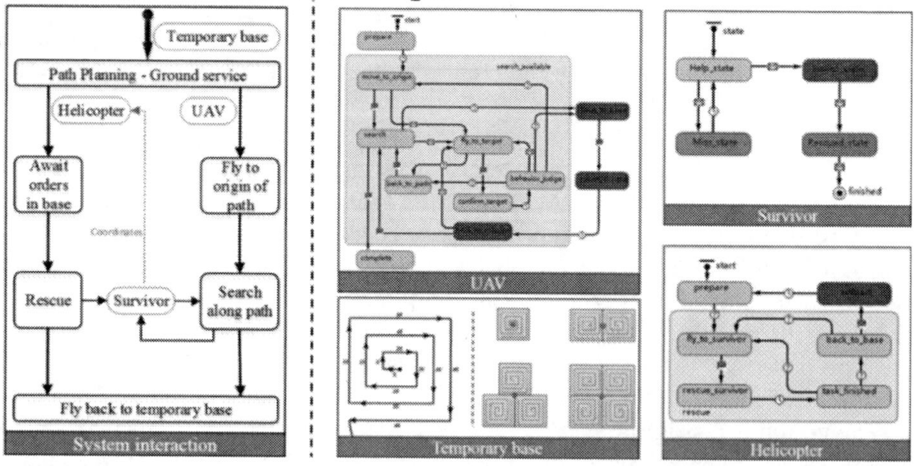

Figure 1. Operation logic of SoS mission.

The operation logic are shown as Fig.1. In the initial condition, survivors are distributed in the search area randomly. The temporary base plans paths based on the number of UAVs and mission area. UAVs carry out their search mission along the search path which is set as an expanding square pattern [12]. When a UAV finds a survivor, it will confirm the coordinates and send message to helicopters. Then a helicopter will fly to target to rescue the survivor. When UAVs finish searching mission area and survivors that are found have been rescued, the mission is finished. Each system has different operation states in its operation logic and systems change states according to active time, environment, coordinates, messages and etc. All logic must be validated through running and debugging the program repeatedly to conform the search and rescue logic in real-life scenario.

2.2 Equations of flight mechanics for UAVs

Without geometric design in early conceptual design phase, it is hard to get some aerodynamics data of UAVs. However, in SoS simulations UAVs operate based on the basic equations of flight mechanics, which provides a bridge from UAV performance to mission effectiveness. That is to say, different original design parameters lead to different UAV performance and different UAV performance lead to different mission effectiveness.

If a UAV is flying at a cruise speed of V, in every moment the motion state of UAVs could be regarded as a steady motion because the total weight or the fuel weight decreases very slowly. Thus the thrust (T) equals the drag (D) and the lift (L) equals the total weight (W). The relationship of the engine thrust (T_a), lift-drag ratio (K), and thrust efficiency (η) can be calculated as Eq. 1. The relationship of W, fuel consumption (Q_f), thrust specific fuel consumption (c_f) and time (t) can be calculated as Eq. 2. The endurance (T_e) and range (R) could be calculated as Eq. 3 and Eq. 4 through the integral from the initial weight (W_1) to the final weight (W_2).

$$T = \eta T_a = \frac{W \times D}{L} = \frac{W}{K} \tag{1}$$

$$dt = \frac{dQ_f}{c_f T_a} = -\frac{dm}{c_f T_a} = -\frac{K\eta dW}{g c_f W} \tag{2}$$

$$T_e = -\int_{W_1}^{W_2} \frac{dW}{g c_f T_a} = \int_{W_2}^{W_1} \frac{K\eta}{g c_f W} dW \tag{3}$$

$$R = -\int_{W_1}^{W_2} \frac{V dW}{g c_f T_a} = \int_{W_2}^{W_1} \frac{VK\eta}{g c_f W} dW \tag{4}$$

From the equations, it can be seen that some performance parameters could be calculated from other parameters, such as T_e and R. So in this simulations, K, W, T/W, c_f, η and Q_f are selected as main design parameters to analyse the UAV performance. However, more design parameters could be added in simulations and analysed in depth as the research continues

3 Analysis Process of SoS simulations

SoS simulations have a significant reference value for aircraft design through exploring design space and analysing mission effectiveness. When operating simulations, the uncertainty must be taken into account. Thus a Monte Carlo simulation is used to capture the stochastic nature in the SoS mission.

3.1 Design space exploration of performance parameters

Before exploring design space, the design target should be decided clearly: a small size, light, portable UAV with a small size piston engine is needed in the earthquake relief. Although designers cannot predict the desirable design point in advance, the bounds of design space could be set based on technical experience or existing UAVs. The bounds should be set wide enough to cover possible design points. The value of parameters are set as a uniform distribution within the bounds to seek possible solution. The value are shown as Table 1.The main structural materials are usually wood, carbon fiber and fiberglass which can make UAVs lighter. So the value of W_{glow} could be set a little small, such as 10 to 15 kg. The thrust of a small size engine is usually low so T/W could be set as 0.4 to 0.8. Compared with low T/W, a high K is needed, such as 12 to 24. c_f and η could be set as 0.8 to 1.2 $kg/(kg \cdot h)$ and 0.7 to 0.9 respectively based on some engine characteristics. And then the space for fuel tank is usually not very large, so W_{fuel} is set as 0.6 to 1.4 kg. Designers could change the bounds according to a certain mission.

Table 1. List of Performance Parameters.

Variable	Description	Distribution
W_{glow}	Gross take-off weight [*kg*]	*uniform*(10,15)
K	Lift-drag ratio	*uniform*(12,24)
T/W	Thrust-weight ratio	*uniform*(0.4,0.8)
c_f	Thrust specific fuel consumption [$kg/(kg \cdot h)$]	*uniform*(0.8,1.2)
η	Thrust efficiency	*uniform*(0.7,0.9)
W_{fuel}	Fuel weight [*kg*]	*uniform*(0.6,1.4)

The simulation scenario is shown as Fig. 2. The green house represents the temporary base. The pink circle represents a helicopter. The light blue circle represents a UAV and the area it has searched is shown as red shadows. The yellow circle represents survivors. They will change their initial coordinates in each run. If found by a UAV, their color changes to red and if rescued by a helicopter, its color changes to blue. In this scenario, an assumption is set that the UAV never miss a survivor. When operating the simulation based on a Monte Carlo, all other mission inputs keep the same except the performance parameters in design space.

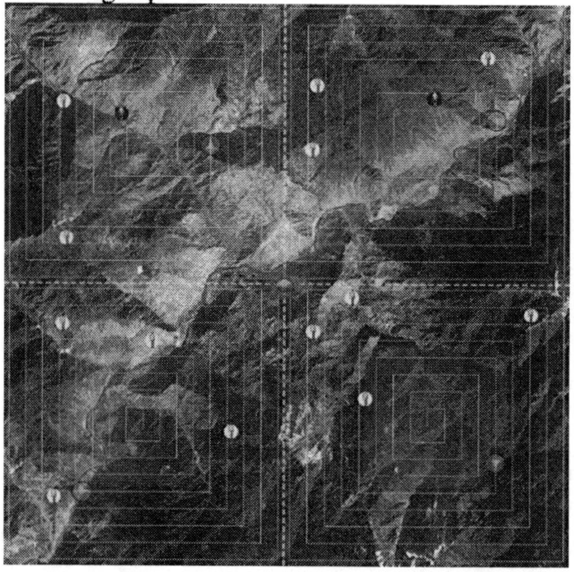

Figure 2. Simulation scenario of SoS mission.

In order to capture the stochastic nature, 30000 design points are used in a Monte Carlo. This amount is accuracy enough to populate the design space after a comparison experiment with 20000 and 40000 design points. In each run, mission time is chosen as the main effectiveness. The data shows that mission time of 30000 cases are all less than 600 minutes. However, the target is to design a competitive UAV and designers are more interested in decreasing the performance parameter without losing much mission effectiveness from the perspective of saving life-cycle cost, too short or too long mission time is not suitable. So the time interval is set as 120 minutes to 180 minutes and the number of cases in this interval is 20656. The histogram and some statistics data are shown as Fig. 3. The histogram in yellow represents the selected cases.

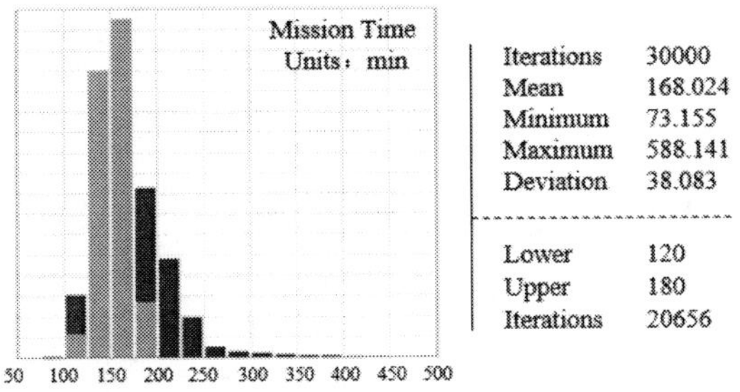

Figure 3. Histogram of All cases and selected cases.

Fig. 3 is not enough to reflect the relationship between performance parameters and mission effectiveness, the data analysis for design parameters should be researched in depth. Here choose *L/D* and *T/W* as an example to illuminate the issue, see Fig. 4. The *x* axis show the interval of parameters. The histogram in light blue represents the distribution of all 30000 cases. The distribution of all cases is uniform because the distribution of the value of design parameters is uniform. The histogram in blue represents the distribution of selected cases. Fig. 4 (a) shows the relationship between *K* and mission effectiveness. It can be seen that the height of blue bars increases sharply when *K* changes from 12 to 18. And then it becomes steady. It is because increasing *K* will decrease the thrust required, which can reduce fuel consumption and improve the endurance and range. The appropriate value could be set as 19 to 20, because it doesn't improve mission effectiveness further with a higher *K*. However, from Fig. 4 (b), it can be seen that the height of blue bars keeps relatively steady in the whole value interval. It is because with a relatively light total weight and a relatively high *K*, *T/W* is not required too high. The appropriate value could be set as 0.4 or 0.5. These optimal parameters will have an important reference value for the following design phases.

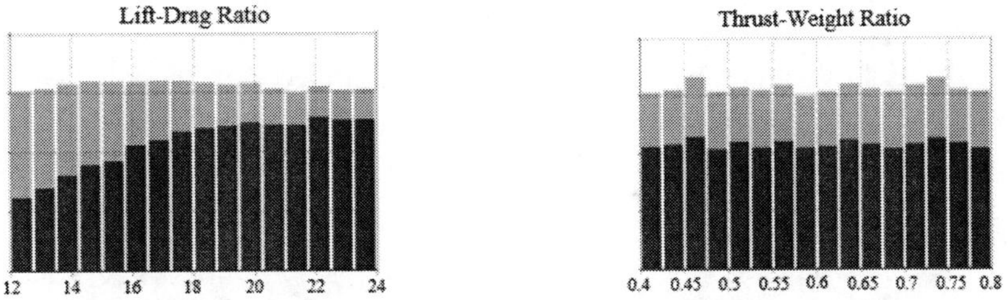

(a). Distribution of *K*. (b). Distribution of *T/W*.

Figure. 4. Histograms of cases from the view of parameters.

3.2 Mission effectiveness analysis of performance designs

If two designs are alternative to carry out a mission, SoS simulations can help designers select the more suitable one from the perspective of overall SoS optimum. The performance parameters are shown as Table 2. When comparing two different UAV designs, another parameter, detection probability (P_{lim}) should be set. It is possible that the UAV with a higher speed may miss more survivors than the one with a lower speed. However, the former one could finish mission faster. So, it is hard to distinguish which one is better in advance before mission simulations. Here mission completeness (D_{mc}) and time utilization ratio (P_{tu}) is set as mission effectiveness, see Eq. 5 and Eq. 6. D_{mc} equals the number of the survivors that are rescued (N_{res}) over the number of all survivors (N_{sur}). P_{tu} equals N_{res} over mission time (T_{fin}). Due to the uncertainty, the mission effectiveness will still change in each run even with the same UAV. Thus a Monte Carlo simulation is also used to capture the stochastic nature. For each UAV, the mission simulation will repeat 2000 times.

$$D_{mc} = \frac{N_{res}}{N_{sur}} \quad (5)$$

$$P_{tu} = \frac{N_{res}}{T_{fin}} \quad (6)$$

Table 2. Performance parameters of two UAV designs.

Variable	Description	UAV A	UAV B
W_{glow}	Gross take-off weight [*kg*]	10	13
K	Lift-drag ratio	16	20
T/W	Thrust-weight ratio	0.4	0.5
c_f	Thrust specific fuel consumption [*kg/(kg·h)*]	0.9	1.0
η	Thrust efficiency	0.7	0.8
W_{fuel}	Fuel weight [*kg*]	1.0	1.2
V	Cruise speed [*km/h*]	80	90
P_{lim}	Detection probability	0.7	0.6

From the view of UAV performance, B is superior to A as Table 2 shows. However, the simulation data show some different results, see Fig. 5. It can be seen from Fig. 5 (a) and (c) that the mean D_{mc} of A is 0.681, higher than B of which D_{mc} is only 0.605. It is because a slow speed can lead to a more accurate research. But the deviation of A is larger than B, because the low K, W_{fuel} and η will lead to a shorter range and endurance. Thus it is easier for A to be affected by uncertainty in the mission. But from the number of cases in which D_{mc} is above 0.5, 0.6, 0.7, 0.8 or 0.9, A is still larger than B. So from the view of D_{mc}, A is superior than B.

From Fig. 5 (b) and (d), it can be seen that there is not much difference in P_{tu}, although the cruise speed of UAV B is higher than UAV A. It is because the time saved by a high cruise speed doesn't play an important role in the whole mission. According to above analysis, A is superior to B. Mission effectiveness analysis based on SoS simulations has an important reference for design evaluation or making mission decisions.

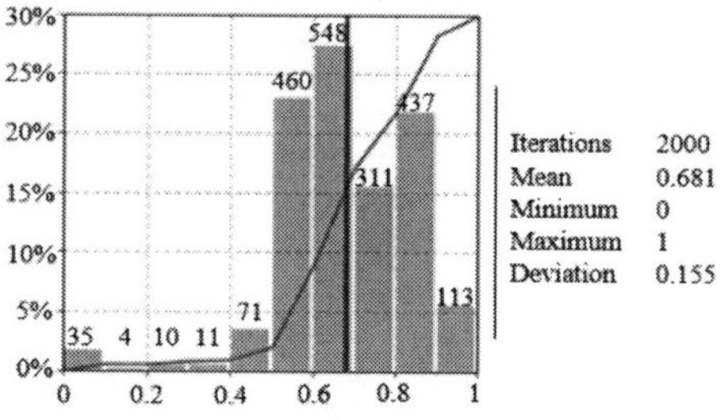

(a). Mission completeness of UAV A.

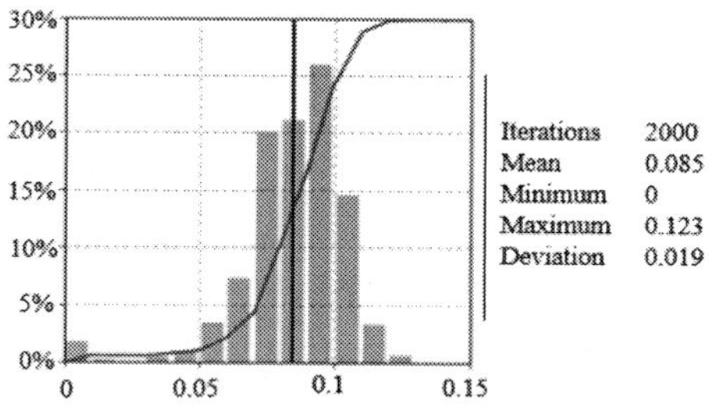

(b). Time utilization ratio of UAV A.

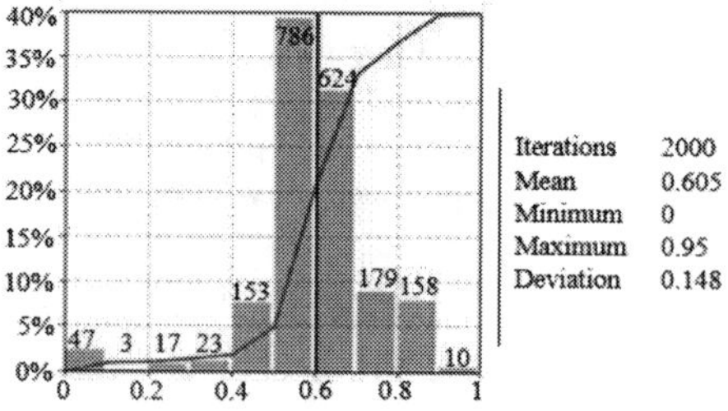

(c). Mission completeness of UAV B.

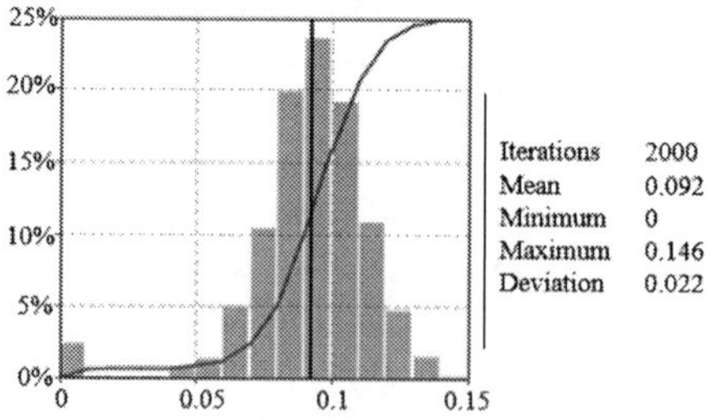

(d). Time utilization ratio of UAV B.

Figure. 5. SoS effectiveness of two UAVs.

4 Conclusion

From the SoS example, it can be seen that SoS simulations provide a feasible way to connect aircraft performance and mission effectiveness. It helps designers to analyse aircraft performance, optimize design parameters and compare different performance designs from the perspective of SoS effectiveness. The results show if designers want to obtain an overall optimum, it may be unnecessary to design aircraft with maximum performance. Through mission simulations, designers could find a more suitable design to save life-cycle time and cost. However, this method still needs more research, for example, more design parameters should be added to simulations as design process continues and design cost should also be taken into account.

References

1. D. Gorissen, E. Quaranta, M. Ferraro, et al, J. AIRCRAFT **51 5**, 1360 (2014)
2. E.S.K. Yu, *Third IEEE International Symposium on Requirements Engineering* (1997)
3. B. Schumann, J. Scanlan, K. Takeda, *Proceedings of the 3rd International Conference on Advances in System Simulation* (2011)
4. K.E. Boulding, MANAGE. SCI. **2 3**, 197 (1956)
5. A. Gorod, B. Sauser, J. Boardman, IEEE SYST. J. **2 4**, 484 (2008)
6. P.T. Biltgen, D.N. Mavris, *45th AIAA Aerospace Sciences Meeting and Exhibit* (2007)
7. D.N. Talley, D.N. Mavris, *49th AIAA/ASME/ASCE/AHS/ASC Structures, Structural Dynamics, and Materials Conference* (2008)
8. M. Mane, W.A. Crossley, A. Nusawardhana, J. AIRCRAFT **44 4**, 1222 (2007)
9. D. DeLaurentis, S. Ayyalasomayajula, J. IND. ECOL. **13 2**, 247 (2009)
10. P. Ranque, D. Freeman, K. Kernstine, D. Lim, E. Garcia, D. Mavris, *11th AIAA Aviation Technology, Integration, and Operations (ATIO) Conference* (2011)
11. M.W. Maier, SYSTEMS ENG. **1 4**, 267 (1998)
12. B. Schumann, J. Scanlan, K. Takeda, *2nd International Air Transport and Operations Symposium* (2011)

MATEC Web of Conferences **52**, 01002 (2016)
DOI: 10.1051/matecconf/20165201002

Combined Cycle Fatigue Investigation Based on Energy Principle

Mykyta Kalynenko[1,a]

[1]*School of Energy and Power Engineering, Beihang University, Beijing 100191, China*

Abstract. We present a modified energy-principle based model of fatigue damage accumulation in high temperature alloys usually used in gas turbine engine under combined high cycle fatigue and low cycle fatigue (LCF/HCF) loading conditions. Our model is based on the energy principle which includes a modified approximation formula that describes fatigue crack origin depending on the relative amplitude of stress intensity in the ranges of both high- and low-cycle fatigue under non-isothermal loading. Functional dependence that presents the influence of HCF mechanisms on a fatigue life of our structural material is gradual and it has not breaks of the curve that yields a possibility to rewrite the equation of the S-N curve with taking into account combined cycle fatigue loading. We used the same number of parameters as the initial model. Note, that new parameter interpretation gives clear physical picture. The proposed model is verified by comparing the computed results with the experimental data for one high temperature alloy GH4133.

1 Introduction

The regime named as a combined cycle fatigue (CCF) is one of the main challenges of the modern fatigue damage analysis and its analysis is a rather complex problem. Thus, in spite of a number of life-prediction methods that have been proposed, their improvements are still discussed intensively. The problem of adequate and full description of accumulation damages under the combined cycle fatigue (low and high cycle fatigue) is rather complicated. The existing widely used fatigue life prediction methods are not correctly representing the fatigue damage accumulation. Often practically used conditions, i.e. elastic stress at high strain amplitudes as an input parameter for the fatigue curve, does take into account the relationship between stress and strain (the real characteristics of the test material, the process of characterizing the cyclic elastic-plastic deformation). On the other hand, approaches based on the amplitude of plastic deformation or plastic deformation work per cycle of loading, no longer work in the transition region between LCF and HCF.

One of the methods which seems is very promising and fruitful is the energy-based life prediction approach. Developed last decades this approach uses clear physical picture of crack nucleation both for low-cycle and high cycle regions of fatigue life. It is assumed that the limit state of a material is determined by that value of the mechanical energy which was dissipated irreversible at cycle deformation. Note, cumulative damage theories based on strain energy were mainly developed in the last several decades and some of the energy-based damage parameters have been proposed.We would

[a]Corresponding author: nkalinenko@gmail.com

like to discuss one of new methods those attempt to combine LCF and HCF and develop CCF approach based on the special representation of the specific energy of cycle deformation as a damage parameter. The approach is based on works of Ellyin and Kujawski [1], , Troshenko and Fomichev [2], Fatemi and Yang [3], Fomichev [4,5]. The approach developed by these authors allows evaluating the strength and lifetime of gas turbine blades by using mathematical models of damaged medium consisting of elastoplastic constitutive equations, kinetic equations of damage accumulation, and a strength criterion of damaged material. It was shown that under regular cyclic loading, when the amplitude of deformation in a cycle decreases, the transition from low cycle fatigue to high cycle fatigue proceeds gradually and depends on the physical interaction of these mechanisms in the transition zone. We suggest some development of theory and our model includes the combined effects of HCF and LCF loads and it is based on the energy-based approach to the analysis of fatigue failure and "energy" criterion for fatigue failure.The fatigue model developed by us expands upon the works of Bol'shukhin et al. (2010) and Volkov et al. (2012). Let us consider the energy dissipation per cycle. In the energy-based approach it is assumed that the limit state of a material is determined by that value of the mechanical energy which was dissipated irreversible at cycle deformation. The fatigue model developed by us expands upon the works of Bol'shukhin et al. (2010) and Volkov et al. (2012). Let us consider the energy dissipation per cycle. In the energy-based approach it is assumed that the limit state of a material is determined by that value of the mechanical energy which was dissipated irreversible at cycle deformation. Some part of this energy will be spending on the fatigue crack origin. As shown below, we propose a modified formula for the function that describes the mechanism of influence of cycle loading on a fatigue life of our structural material.

2 Combined Cycle Fatigue Model

Below we would like to discussed one of new methods those attempt to combine LCF and HCF and develop CCF approach based on the special representation of the specific energy of cycle deformation as a damage parameter. It's considering the energy dissipation per cycle. It is supposed the summary energy dissipated per cycle can be divided on two parts, i.e. i) the so-called "dangerous" part which leads to the fatigue crack origin and ii) a "non-dangerous" part which does not be spent on fatigue failure. Thus, the "dangerous" part the summary strain energy, W_{df},could be written as the difference between total strain energy, W_t, and , its "non-dangerous" part, W_n, see Troshenko and Fomichev [2],

$$W_{df} = W_t - W_n \tag{1}$$

Then, the fatigue failure criterion can be written in the following manner

$$W_{df} = const \tag{2}$$

There are many phenomenological formulas have been proposed to describe the behavior of W_{df}. As it was pointed by Troshenko [6], the analysis of experimental data for a wide range of constructional materials demonstrates that a good result (i.e. when W_{df} is constant) provides the following equation

$$W_{df} = \sum_1^{N_f} \left[\Delta W_{st} - \Delta W_{sc} \left(\frac{\Delta W_{st}}{\Delta W_{sc}}\right)^{\alpha}\right] \tag{3}$$

Here W_{df} corresponds to the fatigue crack origin in the given volume of a metal after N_f cycles, ΔW_{sc} is a specific work of the stress deviators per cycle which is corresponded to the fatigue limit, ΔW_{st} is the total specific work per cycle, α is a parameter of the given material.

In the case of symmetric regular cyclic loading following works we write the fatigue failure criterion as

$$W_{df} = \sum_1^{N_f} \Delta W_{sf} = const \tag{4}$$

where W_{df} corresponds to the fatigue crack origin in the given volume of a metal after N_f cycles.

The total specific work of the stress deviators on the deviators of elastic strain accumulated after N_f cycles is given

$$W_t = \sum_1^{N_f} \Delta W_{st} = \sum_1^{N_f} \sigma_{ij}' \Delta e_{ij}'^e . \tag{5}$$

The "non-dangerous" part W_n is given by the following formulae, see Troshenko and Fomichev [2],

$$W_n = \sum_1^{N_f} \Delta W_{sc} \left(\frac{\Delta W_{st}}{\Delta W_{sc}}\right)^{\alpha} \tag{6}$$

Here ΔW_{sc} is specific work of the stress deviators per cycle which is corresponded to the fatigue limit. Thus, combining these formulas we have the following fatigue failure criterion (comp. Eq. (4))

$$\sum_1^{N_f} \Delta W_{sf} = \sum_1^{N_f} \left[\Delta W_{st} - \Delta W_{sc} \left(\frac{\Delta W_{st}}{\Delta W_{sc}}\right)^{\alpha}\right] = const \tag{7}$$

In the case of irregular cyclic loading we can rewrite Eq. (3) for the loading step $\Delta t = \mathrm{t}_{n+1} - \mathrm{t}_n$, in following form, see Volkov and Korotkikh [7],

$$\Delta W_0 = \Delta W_e (1 - f(\gamma)) \tag{8}$$

where $\gamma = \sigma_u / \sigma_{uy}, \Delta W_e = \sigma_{ij}' \Delta e_{ij}'^e / 2$; $\sigma_u = \sqrt{\sigma_{ij}' \sigma_{ij}'}$ is the intensity of stress tensors; $\Delta e_{ue} = \sqrt{e_{ij}'^e e_{ij}'^e}$ is the incrementing intensity of plastic deformation, σ_{uy} is the intensity of stress tensor which is correspondent to endurance limit stress; ΔW_0 is the "dangerous" part of the specific energyΔW_e; the "non-dangerous" part under the loading is written as $\Delta W_e f(\gamma)$.

Function $f(\gamma)$ describes the mechanism of influence of HCF on a fatigue life of our structural material. Here we would like to discuss some approximation of the function $f(\gamma)$. Usually (see e.g. Bol'shukhin [8]), the following approximation is used

$$f(\gamma) = \begin{cases} 1 & if \gamma \le 1 \\ 1 - b^* \left(\frac{\gamma - 1}{\gamma^* - 1}\right)^m & if \ 1 \le \gamma \le \gamma^* \\ 1 - b^* & if \gamma \ge \gamma^* \end{cases} \tag{9}$$

Here b^* is an asymptotic value of the $f(\gamma)$ at $\gamma \to \gamma^*, \sigma_u \to \sigma_u^*$ and $0 < b^* < 1$. Fig. 1 represents the expected behavior $f(\gamma)$ as a function of a variable γ.

Generally speaking, this function $f$depends on four variables but three of them, i.e. b^*, γ^*, m, are considered as parameters. The values of these parameters should be determined from experimental data. Here we should note the curve that presented at Fig. 1, has a convex type in the region $\gamma \in [1, \gamma^*]$. It is clear that this convexity is determined by the following demand$m \ge 1$. It is the case that experimentally supported and discussed in literature last time (see, e.g. [8-10]).

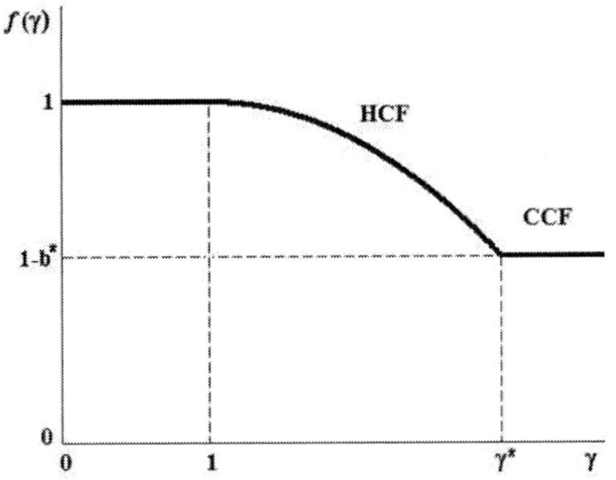

Figure 1. Function f that describes the mechanism of influence of HCF on a fatigue life of a structural material vs the normalized intensity of stress tensorsγ.

Let us discuss the physical picture of material behavior that is described by this model of the function $f(\gamma)$. There is no expected the appearing of a fatigue crack in the interval $\gamma \in [0,1]$. But, fatigue cracks may appear due to HCF mechanisms in the region $\gamma \in [1,\gamma^*]$. At last, mutual act of combined HCF and LCF mechanisms of a fatigue crack origin is expected in the region where $\gamma > \gamma^*$.

It is seems clear, that such hard separation of the regions is quite approximate in real. We propose here to use another equation for the function $f(\gamma)$ in hope that it is more gradual. Thus, it gives us possibility to write some analytical formulas.

Let us write the following equation

$$f_n(\gamma, b^*, \mu, T_m) = \frac{1-b^*+e^{\frac{\mu-\gamma}{T_m}}}{1+e^{\frac{\mu-\gamma}{T_m}}} \tag{10}$$

Our function, $f_{n,}(\gamma)$ depends on three additional parameters too. We use for them the following symbols: b^*, μ and T. The functional dependence is a modified version of a logistic function which is the well-known Fermi-Dirac distribution. It is supposed that T_m is small as to compare with μ. Parameters μ and T_m in this case have a well-defined meaning. Thus, μ is the middle point of the transitional region of the width T_m. Meanwhile, Eq. (10) gives more gradual approximation than the Eq. (9), see Fig. 2.

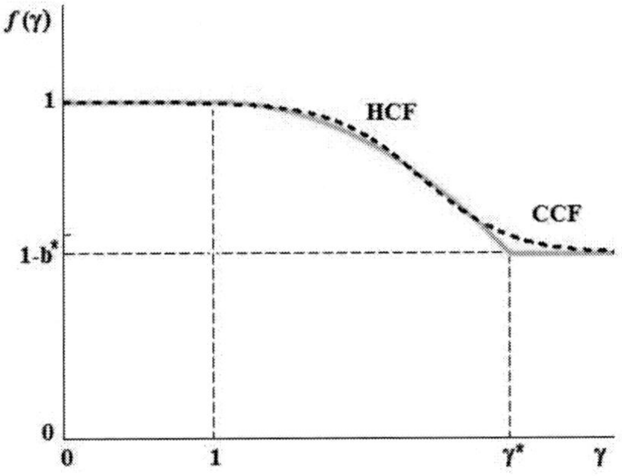

Figure 2. Comparison between functional dependences of the modifiedfunction $f_n(\gamma)$ (dashed line) and an usual approximation of the function $f(y)$ (solid line) at the same value of the parameter b^*.

Note, that in our model there is no sharp boundary between the region where fatigue cracks are caused by the HCF mechanisms and the region where CCF mechanisms is expected. It seems to us, that this physical picture is quite realistic as to compare with the sharp change of the fatigue crack origin mechanism. Then, parameter b^* has the same physical meaning that it is in the model which is described by Eq. (9). Namely, it bound the minimal value of the "non-dangerous" part of the specific energy under the loading. But now, the process of fatigue crack appearing is described by the smooth transition from HCF to CCF mechanisms. The "middle point" of the transition between HCF and CCF processes is described now by the parameter μ. The parameter T_m specifies the width and sharpness of the transition curve.

Let us look what are the consequences. Following [8-10] we write the structure of the evolution equation of damage accumulation for the HCF region in a manner like the LCF region (for details see Mitenkov et. al. [10]). Integrating this equation for the given case of loading process give us the equation which bounds the total damage accumulation in an elementary volume, ω, with the total relative specific "dangerous" energy, z_s, that has been accumulated as a result of loading.

For the sake of simplicity, we limit ourselves here the following model. Let the function $F_v(\beta)$, that describes the influence of stress conditions to the process of damage accumulation, is constant. Then, we may write the following equation

$$\omega = 1 - [1 - F_v(\beta) z_s^{\alpha+1}]^{1/(r+1)} \tag{11}$$

Firstly, we discuss the case of cyclic symmetrical loading of a cylindrical sample at uniaxial tensile and compression test with the amplitudeof the intensity of stress tensors σ_{ua} (which is corresponded to the intensity of the strain tensor $e_{ua} = \sigma_{ua}/(2G)$, where$G$is the shear modulus). The specific energy per cycle be written

$$\Delta W_{st} = \frac{e_{ua}\,\sigma_{ua}}{2} \cdot 4 = \frac{\sigma_{ua}}{G} \tag{12}$$

The "dangerous" part of the specific energy per cycle, ΔW_{sf}, is given by

$$\Delta W_{sf} = \Delta W_{st}\left(1 - f_n(\gamma_a)\right) = \frac{\sigma_{ua}^2}{G} \cdot \frac{b^*}{1+e^{\frac{\mu-\gamma_a}{T_m}}} \tag{13}$$

Here $\gamma_a = \sigma_{ua}/\sigma_{cu}$, σ_{cu} is the intensity of stress tensor which is corresponded to the fatigue limit The relative specific "dangerous" energy per cycle is given by (compare with Eq. (12) in paper of Bol'shukhin et al. [8])

$$\Delta z_{st} = \frac{\Delta W_{st}}{\Delta W_{fe}}\left(1 - f_n(\gamma_a)\right) = \frac{\sigma_{ua}^2}{GW_{fe}} \cdot \frac{b^*}{1+e^{\frac{\mu-\gamma_a}{T_m}}} \tag{14}$$

Here W_{fe} is the critical value of the specific "dangerous" energy in HCF regime.

Thus, the total relative specific "dangerous" energy which has been accumulated after Ncycles is given by

$$z_s = \sum_1^N \Delta z_{st} = \frac{\sigma_{ua}^2}{GW_{fe}} \cdot \frac{b^*}{1+e^{\frac{\mu-\gamma_a}{T_m}}} \cdot N \tag{15}$$

The total damage accumulation, ω_f, after Ncycles, as it was written in [1], we obtain in the following form

$$\omega_f = 1 - [1 - F_v(\beta) z_s^{\alpha+1}]^{1/(r+1)} \tag{16}$$

where the values of parameters should be determined by fitting experimental data, $\beta = \sigma/\sigma_u$, $\sigma_u = \sqrt{\sigma'_{ij}\sigma'_{ij}}$, $\sigma = \sigma_{ii}/3$. At $N = N_f$ a fatigue crack appears and we have $\omega_f(N_f) = 1$.

As it follows from Eqs. (15-16), we obtain the following equation for the S-N curve

$$N_f = \left(\frac{1}{F_{vm}(\beta)}\right)^{\frac{1}{1+\alpha}} \cdot \frac{GW_{fe}}{\sigma_{ua}^2} \cdot \frac{1}{b^*} \cdot \left(1 + e^{\frac{\mu-\gamma_a}{T_m}}\right) \tag{17}$$

Here $F_{vm}(\beta)$ is the middle value of our function F_v per cycle: $F_{vm}(\beta) = \left(F_v(\beta_c) + F_v(\beta_p)\right)/2$, $\beta_c = \sigma_t/\sigma_{ua}$, $\beta_{pc} = \sigma_c/\sigma_{ua}$.

Let's introduce a new constant B_1

$$B_1 = \left(\frac{1}{F_{vm}(\beta)}\right)^{\frac{1}{1+\alpha}} \frac{GW_{fe}}{b^*} \tag{18}$$

Then Eq. (17) we rewrite in the following manner

$$N_f = B_1 \frac{1+\exp\left\{\frac{\mu - \left(\frac{\sigma_{ua}}{\sigma_{cu}}\right)}{T_m}\right\}}{\sigma_{ua}^2} \tag{19}$$

Generally speaking, the dependence $N_f(\sigma_{ua})$, that is described by Eq. (19) in the presented model, is not so sharp as to compare with the power law of Eq. (27) in the paper of Bol'shukhin et al. [8].

Here we should note, that at large values of stresses σ_{ua}, Eq. (19) in the main approximation provides the formulae of the following type

$$2N_f = B\frac{1}{\sigma_{ua}{}^2} \tag{20}$$

where constant B one may calculate with taking into account Eqs. (18) and (19).

However, to compare the model results with experimental data it is better to use Eq. (19). In this case we have to solve it numerically to reveal S-N curve in coordinates σ_{ua} and N_f.

The asymptotic value of the relative specific dangerous energy per cycle, b^*, is given as

$$b^* = \frac{\Delta W_{sf}^*}{\Delta W_{st}} \tag{21}$$

Parameters $\mu$and $T_m$in Eq. (19) should be determined in a such way that function $f_n(\gamma_a, b^*, \mu, T_m)$ possess the value which is equal to $(1 - b^*)$ with the given accuracy to the given material.

Usually, the shift along the axis γ on $4T_m$from the value of μ yields the deviation from the value of $(1 - b^*)$which is of the order of 1%. This value of the $\mu + 4T_m$ should be corresponded to the technological yield stress of the given material.

The developed model is verified by comparing the data of experiments on uni-axial tensile testing for high-temperature GH4133 alloy at 300 °C with the fitted computed fatigue curve. Figure 3 presents the comparison between tests data for high temperature alloy GH4133 at temperature T= 300 °C and data which has been obtained as s result of numerical calculation of our modified model using Eq. (19).

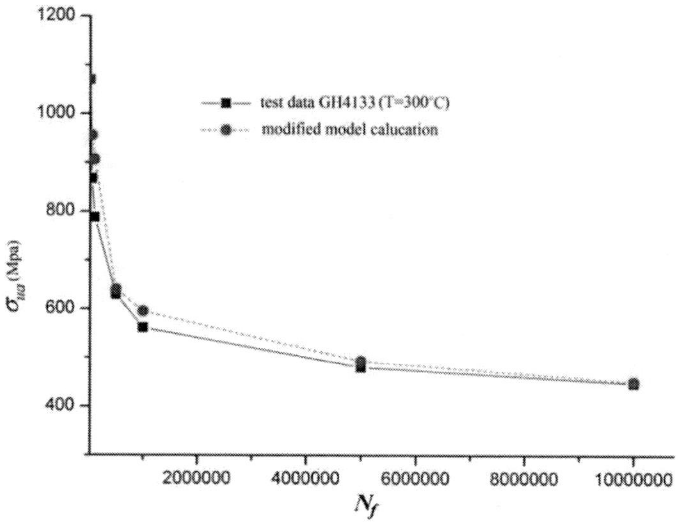

Figure 3. Comparison between curve $\sigma(N)$ predicted by the proposed model (circles) and those tested for GH4133 at 300 °C under symmetric loading (squares).

It can be seen from Fig. 3 that calculated curve is close enough to the experimental results for GH4133 at temperature T= 300 °C. Temperature increasing change slightly this picture.The expanded scope of this approach with comparing results of calculations using different models of CCF will be published shortly.

3 Conclusion

The aim of current research was to present a modified energy-principle based model of fatigue damage accumulation in high temperature alloys usually used in gas turbine engine under combined high cycle fatigue and low cycle fatigue loading conditions. In this paper, model is based on the energy principle which includes a modified approximation formula that describes fatigue crack origin depending on the relative amplitude of stress intensity in the ranges of both high- and low-cycle fatigue under non-isothermal loading. Functional dependence that presents the influence of HCF mechanisms on a fatigue life of our structural material is gradual and it has not breaks of the curve that yields a possibility to rewrite the equation of the S-N curve with taking into account combined cycle fatigue loading. We used the same number of parameters as the initial model. Note, that new parameters specify the width and sharpness of the transition region between HCF and CCF mechanisms.The proposed model is verified by comparing the computed results with the experimental data for one high temperature alloy GH4133.

Acknowledgment

This work is partly supported by National Natural Science Foundation of China (Grant No. 11272025) and Defense Industrial Technology Development Program (Grant No. B 2120132006).The author would like to thank Prof. Xiaojun Yan (Beihang University) for fruitful and helpful discussions.

References

1. K. Ellyin, D. Kujawski, J. “Microstructure and Mechanical Behavior of Materials” EAMS, **11**, pp. 591 (1986)
2. V.T.Troshchenko, P.A.Fomichev, Strength of Materials, **25**, 1 (1993)
3. A.Fatemi, L. Yang, Int. Journal of Fatigue, **20**, 9 (1998)

4. P.A.Fomichev, Strength of Materials, **27**, 355 (1995)
5. P.A. Fomichev, Strength of Materials, **36** , 231 (2004)
6. V.T. Troshchenko, *Deformation and damage of metals under cyclic load*(Naukova dumka, Kiev, 1981)
7. I.A.Volkov, Yu.G.Korotkikh, *Equations of state of viscoelastoplastic damaged media*, (Fizmatgiz, Moscow, 2008)
8. M.A.Bol'shukhin, D.L.Zverev, V.B.Kaydalov, Yu.G.Korotkikh, V.A.Panov, V.A.Pakhomov, Problems of strength and plasticity, 72, 29 (2010)
9. I.A.Volkov, M.N.Ereev, Yu.G.Korotkikh, I.S.Tarasov, Computational Continuum Mechanics, **5**, 54 (2012)
10. F.M. Mitenkov, V.B. Kaidalov, Y.G. Korotkikh, *Methods for justifying the service life of nuclear power plants* (Mashinostroenie, Moscow, 2007)

MATEC Web of Conferences **52**, 01003 (2016)
DOI: 10.1051/matecconf/20165201003

Circuit Modeling and EM Simulation Verification of DGS based Low-Pass Filter Employing Transmission Line Model along with Microstrip-Slotline Transitions

Mouloud CHALLAL[1,a], Mokrane DEHMAS[1], Arab AZRAR[1], Rabia AKSAS[2] and Mohamed TRABELSI[2]

[1]*Signals and Systems Laboratory, Institute of Electrical and Electronic Engineering, Dept. of Electronics, University of Boumerdes, 35000 Boumerdes, Algeria*
[2]*Electronics Department, Ecole Nationale Polytechnique, 16000 Algiers, Algeria*

Abstract. In this paper, an equivalent circuit model (ECM) for a defected ground structure (DGS) pattern is proposed and evaluated for designing a compact low-pass filter (LPF). The proposed ECM is based on microstrip lines and microstrip-slotline transitions. Every slotlines of the DGS unit are modeled by ideals transmission lines of characteristic impedance and electrical length. Comparison between full-wave EM and circuit simulations illustrates the validity of the proposed ECM.

1 Introduction

Recently, various etched defects on metallic ground plane, namely, defected ground structure (DGS) have been focused by many researchers. They are very efficiently used to improve the performances of a variety of microwave circuits such as filters, amplifiers, oscillators and antennas [1]–[4]. Usually, a DGS unit in either microstrip or coplanar waveguide is modeled by lumped elements (LE), i.e, a parallel LC resonant circuit, known as conventional equivalent circuit model (CECM) connected to transmission lines at both sides [1]. Furthermore, some circuit models based on an inductor and a shorted stub for a spiral DGS and a short stub, additional series transmission lines and a resistor for a dumbbell DGS have also been presented in literature [5], [6]. Even though this type of circuit models can present a periodic frequency response, they require a full-wave electromagnetic (EM) simulation in order to determine the values of their elements. However, EM simulation takes a lot of time to accomplish the design goals. This is a first drawback of this modeling method. The second one is that there is not relationship between the circuit model parameters and the physical dimensions of the DGS. Consequently, this modeling technique is not practical for designing microwave circuits in general and filters in particular. To overcome these limitations, it is necessary to establish an accurate technique by developing another equivalent circuit model (ECM), which is derived from DGS physical dimensions and do not need to use EM simulation to achieve it parameters.

In this paper, a proposed equivalent circuit model (PECM) for the DGS is developed and a systematic method for the parameters extraction is described. Moreover, it is applied to the design a compact low-pass filter (LPF). The PECM is based on microstrip line model and microstrip-slotline transitions [7]-[9]. Every slotlines of the DGS are modeled by ideals transmission lines of

[a] Corresponding author : mchallal@ieee.org

characteristic impedance (Z_{Si}) and electrical length (θ_{Si}), where i goes from 1 to nb (nb is the slotlines number). To show the validity of the PECM, three steps are taken in consideration and are as follows:

(i) Comparison between Full-wave EM, CECM and PECM simulations of the DGS unit;

(ii) Application of the PECM to design a compact LPF having three DGS units as an example of purpose;

(iii) Comparison between full-wave EM-LPF and PECM-LPF simulations.

The availability of substrate material, which is used in this work, is based on RO4350B Rogers material with a thickness of 0.25 mm, conductor layers of 35 μm and a relative permittivity of 3.63.

2 Equivalent Circuit Model

Based on microstrip line model and microstrip-slotline transitions [7]-[9], a PECM is described in this section. As a first step, a dumbbell shaped DGS is chosen as an example of application that is shown in Figure 1.a. Next, the structure is modeled by a conventional parallel inductor (L) and capacitor (C) resonant circuit connected to transmission lines at both sides as shown in Figure 1.b, where the circuit elements are given by [1].

$$C = \frac{\omega_0}{2Z_0(\omega_0^2 - \omega_c^2)} \tag{1}$$

$$L = \frac{1}{\omega_0^2 C} \tag{2}$$

where $\omega_0 = 2\pi f_0$ and $\omega_c = 2\pi f_c$ are respectively the angular resonant frequency and 3-dB cutoff frequency (f_0 and f_c are respectively the attenuation pole and cutoff frequencies, they obtained from the simulated response of a DGS unit using an EM simulator) of the DGS. Z_0 is the characteristic impedance of the transmission line.

The considered DGS unit is composed of two rectangular defected areas and a narrow connecting slot as shown in Figure 1.a. The dimensions a, b, g and c are considered to be respectively 5, 3.8, 0.5 and 1 mm. Hence, the achieved attenuation pole and cut-off frequencies using an EM simulator are respectively 8 GHz and 3.68 GHz. Insertion of these quantities into equations (1) and (2) yield values of C and L to be respectively 0.25 pF and 1.57 nH. A comparison between CECM, PECM and full-wave EM simulations will be discussed later.

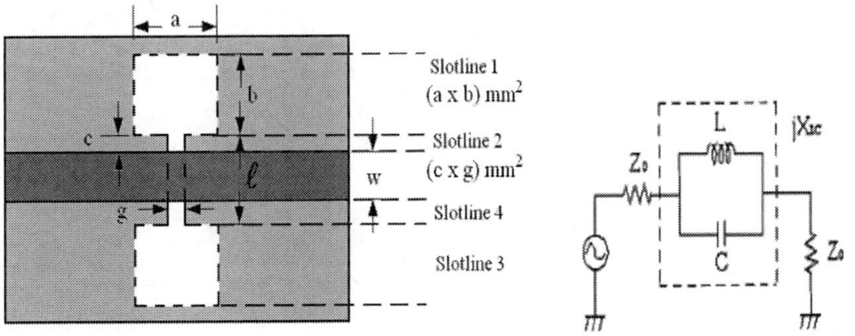

Figure 1. DGS unit. (a) Geometry of dumbbell shaped DGS and, (b) Its conventional equivalent circuit model (CECM)

The PECM for the DGS unit is shown in Figure 2 (Model 1 which corresponds to one DGS unit). The considered structure is modeled such that, all slotline are modeled by ideal transmission lines of characteristic impedance (Z_{Si}) and electrical length (θ_{Si}), where i goes from 1 to nb (nb is the slotlines number) as shown in Figure 2. In our case, the Slotlines 1 and 3 and Slotlines 2 and 4 indicated in

Figure 1.a are respectively the two rectangular defected areas (axb) and a narrow connecting slot (gxc). Next, the short-end of the slotline is represented by an equivalent length [9].

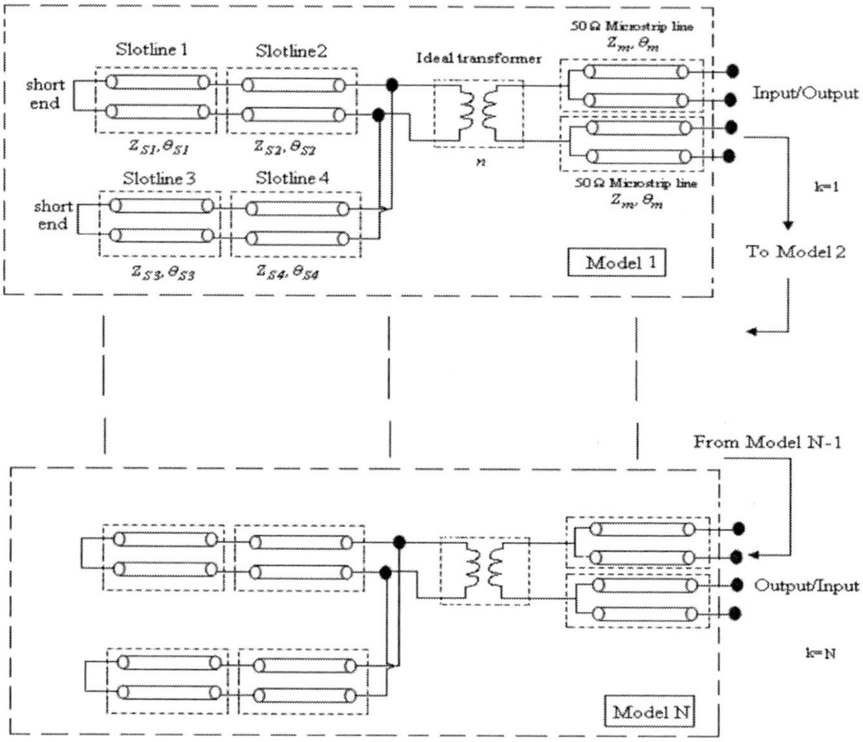

Figure 2. Proposed equivalent circuit model (PECM) for the cascaded defected ground structures (N: Number of used DGS patterns)

Finally, the transition of microstrip line to slotline, i.e, the EM coupling between them is modeled by an ideal transformer as shown in Figure 2, where n is the turn ratio of the transformer. In this work, for the simplification raison of calculus, the slotlines 1 and 2 are considered respectively similar to slotlines 3 and 4 as shown in Figure 1.a. The factor k of Figure 2 is equal to the number of similar DGS patterns connected in cascade.

The slotline equations such as slotline wavelength and slotline impedance used in this work are the closed form expressions of Janaswamy and Schaubert [9] for low ε r substrates obtained by curve fitting the numerical results achieved from Galerkin' s method in the Fourier transform domain (FTD). Theoretical and measurements results of the short-end of the slotline obtained by Yang and Alexopoulos [10] and Knorr and Seanz [11], respectively, have expected that the end reactance of a shorted slotline is purely inductive and significant. It increases with the increase of both the slotline width and h/λ_0 ratio. The inductive reactance is expressed by an inductance (L_{0S}) connected to the end of a slotline. The equivalent length of the short-end effect (ℓ_{S1_eq}) may be up to $0.1\,\lambda_{S1}$ [9].

Using Agilent ADS simulator, the PECM is simulated for the purpose of comparison to the CECM and full-wave EM simulations. The schematic circuit of the PECM in ADS simulator is shown in Figure 3. The electrical lengths l_{S1_eq}, l_{S1} and l_{S2} mentioned in Figure 3 correspond respectively to the electrical length of the short-end effect (θ_{S1_eq}), θ_{S1} and θ_{S2}. The values of θ_{S1} and θ_{S2} are selected by considering f_0 of 8 GHz in.

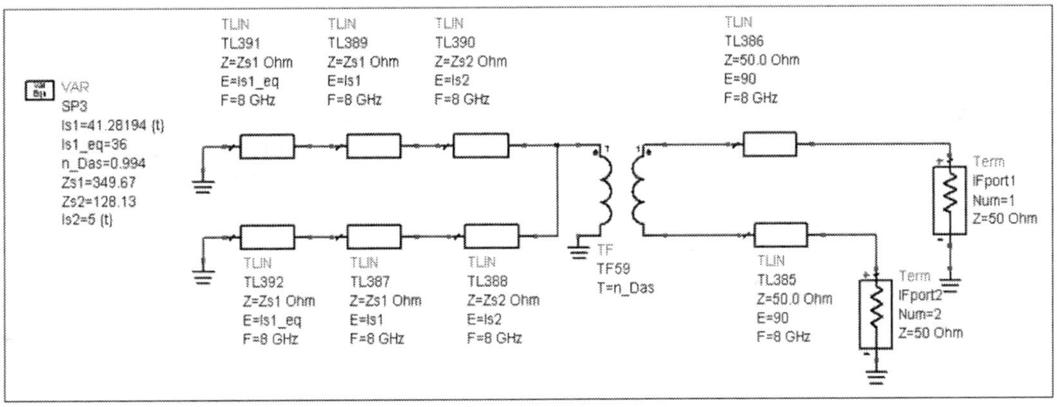

Figure 3. Schematic circuit of the PECM in ADS simulator

The required physical length for the first ideal transmission line including the equivalent length of the short-end effect (L_1) as a function of both ratio λ_{S1}/λ_0 (given in [9] for $0.075 \le a/\lambda_0 \le 1$) and f_0 may be estimated by the following expression:

$$L_1 = L_{S1} + \ell_{S1\_eq} \approx 0.1145 \frac{c\,10^3}{f_0} \frac{\lambda_{S1}}{\lambda_0} + 0.1\,\lambda_{S1} \approx 343.5 \frac{10^8}{f_0} \frac{\lambda_{S1}}{\lambda_0} + 0.1\,\lambda_{S1} \quad (3)$$

where L_{S1} is the physical length of the first ideal transmission without the equivalent length of the short-end. The frequency f_0 is given in GHz.

Similarly to the estimated L_1, the required physical length for the second ideal transmission line as a function of both ratio λ_{S2}/λ_0 (given in [9] for $0.0015 \le g/\lambda_0 \le 0.075$) and f_0 may be estimated by the following expression:

$$L_2 \approx 0.0137 \frac{c\,10^3}{f_0} \frac{\lambda_{S2}}{\lambda_0} \approx 41.1 \frac{10^8}{f_0} \frac{\lambda_{S2}}{\lambda_0} \quad (4)$$

Figure 4 shows comparisons of EM and circuit models simulations. In Figure 4, it is observed that an excellent agreement has been found between Full-wave EM and PECM simulations. This result shows that the PECM provides excellent performances and gives results comparable to the conventional one.

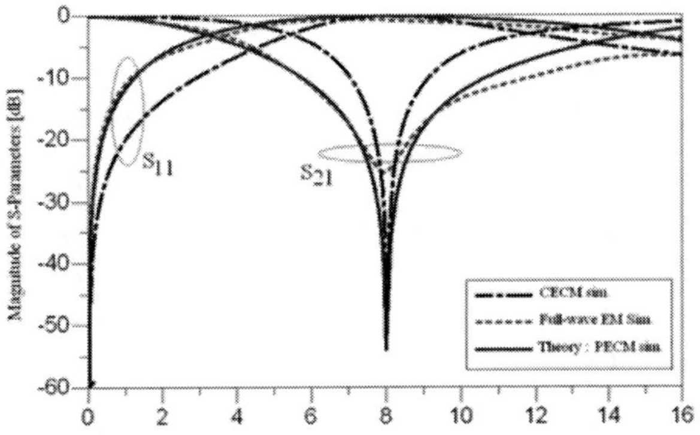

Figure 4. Magnitude of S_{21} and S_{11} of DGS unit, CECM, PECM and Full-wave EM simulations

3 Application of the proposed equivalent circuit model to design a microwave LPF with DGS

In order to show the validity of the PECM, a DGS-dumbbell LPF has been designed and simulated. To improve the transfer function response (S_{21}) of the DGS unit shown in Figure 4, full-wave EM and circuit simulations with increasing number of cascaded dumbbell-shaped DGS units are performed. It has been observed that a good response is achieved when three or more units are cascaded. Figure 5 shows the layout of three units of DGS-LPF whereas Figure 6 shows the magnitude of S_{21} and S_{11} obtained by both full-wave EM and PECM simulations.

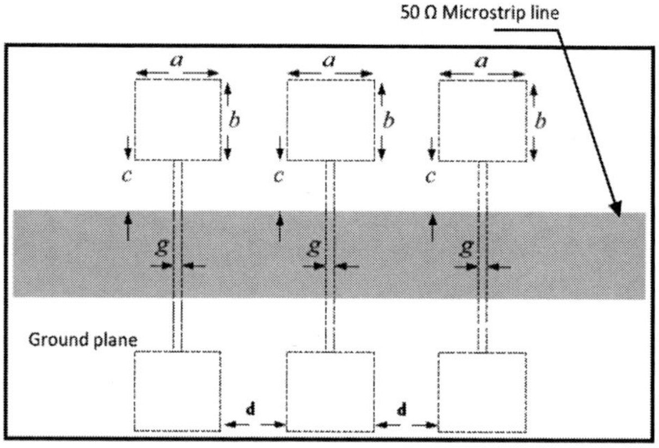

Figure 5. Layout of LPF based on three DGS slots

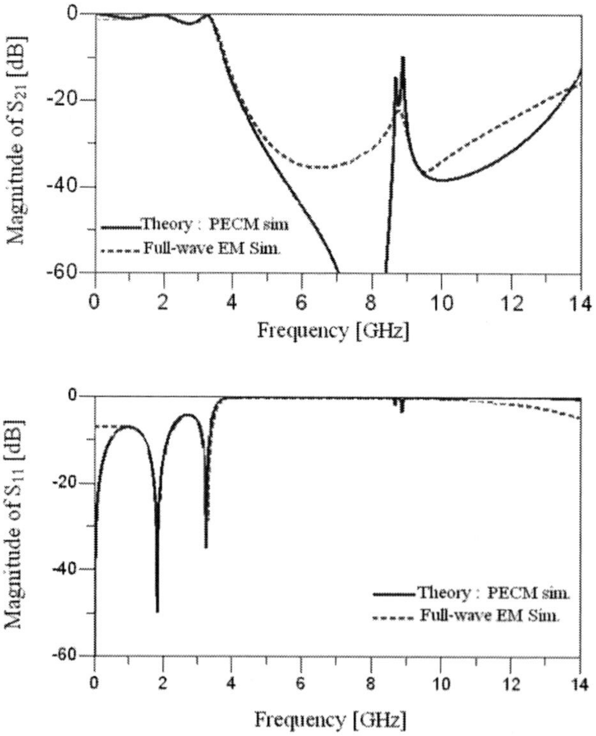

Figure 6. Magnitude responses of the DGS-LPF : Theory (PECM simulation) and full-wave EM simulation

From Figure 6, it can be seen that the simulation results obtained by the PECM (theoretical) agree well with full-wave EM-simulations results. Besides, less time is consumed when simulations are performed using the PECM. Consequently, when the number of DGS in a circuit is large (eg. more than two DGSs), the simulations using the PECM is more advantageous than the full-wave EM simulations.

4 Conclusion

An Equivalent circuit model (ECM) using transmission line model along with microstrip-slotline transitions has been proposed to illustrate and design defected ground structure (DGS) unit. All slotlines of the DGS unit have been modeled by ideals transmission lines of characteristic impedance Zs and electrical length θ_S. Also, The EM coupling between microstrip line and slotline is modeled by an ideal transformer. To qualify the proposed circuit model, a microwave LPF with DGS based on the proposed ECM is designed. A good agreement between the circuit simulation and full-wave EM-simulations of the filter characteristic validates the proposed ECM of the microwave DGS based LPF.

Acknowledgment

The authors express their thanks to Dr. Khaled Ben Ali, University catholique de Louvain, Belgium, for providing support and assistance to perform simulations by using software.

References

1. F. Wei, X. Shi, B. Li, Q. Huang and X. Wang , Micowave and optical technology letters, **50**, (2008).
2. A. Boutejdar, M. Challal and A. Azrar, The Applied Comp. Electromagnetics Society, **26**, (2011)
3. Y. T. Lee, J.S. Lim, J.S. Park, D. Sangwook Nam, IEEE Microwave and Guided Wave Letters, **12**, 2002.
4. M. Challal, A. Azrar and M. Dehmas,, IJCSI- Int. J. of Comp. Science Issues, **8**, (2011).
5. C. S. Kim, J. S. Lim, S. Nam, K. Y. Kang and A. Ahn, Electron. Lett., **38**, (2002).
6. J. Park, K. Park, S. Chang and D. Ahn, Proc. 4th International Conference on Electronics, Hardware, Wireless and Optical Communications, (2005)
7. J. B. Knorr, IEEE Trans. Microwave Theory Tech., (1974)
8. B. Schuppert, IEEE Trans. Microw. Theory Tech., **36**, (1988)
9. K. Gupta, R. Garg, I. Bahl, and P. Bhartia, 2nd ed, Artech House, (1996)
10. H. Y. Yang and N. G. Alexopoulos, IEEE Trans. Microwave Theory Tech., **36**, (1988)
11. J. B. Knorr and J. Seanz, IEEE Trans. Microwave Theory Tech., **21**, (1973)

MATEC Web of Conferences **52**, 01004 (2016)
DOI: 10.1051/matecconf/20165201004

Requirements Engineering Model: Role Based Goal Oriented Model

Sandfreni, S.SI, MT[1, a], Ir. Kridanto Surendro, M.Sc, PhD[2]

[1]*University of Esa Unggul, Faculty of Computer Science, Jakarta, Indonesia*

[2]*Bandung Institute of Technology, School of Electrical Engineering and Informatics, Bandung, Indonesia*

Abstract. Requirements engineering approach through intentional perspective is one of the arguments that appear in the field of requirement engineering. That approach can explain the characteristics of the behavior of an actor. The usage Goal Based Workflow and KAOS method in iStar modeling might help the system analyst to gain knowledge about the internal process inside each of actor sequentially, such that the whole sequential activity to achieve the goal are exposed clearly in those actor's internal process. The adoption of the concept of the role of RACI diagram on Role Based Goal Oriented Model system analyst gain complete knowledge about requirements of actor who involve in a system. System analyst might also distinguish the dependency between each actor in each process. Those dependencies are exhibited in strategic dependency model. In addition, the internal activities of the actor are also shown in strategic rationale model.

1 Introduction

Requirement Engineering is the first step in the adoption of an information system. Requirements Engineering has a purpose to understand the system users needs, which is realized by identifying the involved component in the system. Requirement Engineering in Software Engineering that identifies the actual user's requirement is a critical part that determines the successfulness of system development. The main constraint in gathering the requirement is the difficulties in understanding the application domain [1].

Designing a system that suitable with the user requirement is still challenging. This challenge is lied on the difficulty to rely on the results of mainly found in the difficulty to rely on the results of catching system requirement [2]. When users express their needs, all of those requirements is only the beginning concept of the solution, hence the actual needs is still unrevealed [2]. Because Requirement Engineering is a bridge that connects the world of informal stakeholder needs with the world of formal system behavior, therefore the understanding of the system, the conceptualization of system character, identification of the system constraints, understanding of system development cycle, and system analyzation are required [3].

Requirements engineering approach through intentional perspective is an arguments that appear in the field of requirement engineering. That approach is able to explain the characteristics of an actor behavior, with the expected model by an actor that can be obtained without specifying their behavior

[a] Corresponding author : sandfreni@esaunggul.ac.id

precisely. In the i*(iStar) modeling that focused on intentional perspective and understand the relationship between actors in a domain of behavior, but rather to how the system can help improve the relations between social actors in an organization [4].

Several researches on iStar modeling have produced various requirements engineering model. iStar modeling has represented successfully the relation between actors in term of tasks, goals and resources. iStar can also describe the processes that occur on each actor. iStar has been modified to represent the requirements model in hierarchical based on organizational structure, in such way relationship between actor are possible to be observed on organization level that has been written in *Requirement Engineering Model for Cloud Computing University Using iStar Hierarchy Method* [5]. But until now the development of modeling iStar can't representing about role in a process explicitly.

2 Literature Review

2.1 The Approach of Goal to Understand The Organization's Situation and Needs

In understanding the organization needs, the way to represent knowledge of the domain, structurally is needed. Describe the overall business situation and also includes cooperation between actors and goals of each group in the organization. In describing the situation, goal has a function as a motivation resource from each intentionality, which is manifested in actions, or procedures that occur in the organization.

2.2 Goal Based Workflow

Goal based workflow approach describes the interaction between actor that associated with goal. In this approach goal is represented explicitly. The model express the organization of tuple [G, A, R], where G is the "goal" set, A is the "actor" set, and R is the "resource" set [6]. A set of actor will collaborate utilizing a set a resource to achieve the organization's goal from an organization, such as depicted in the figure 2.

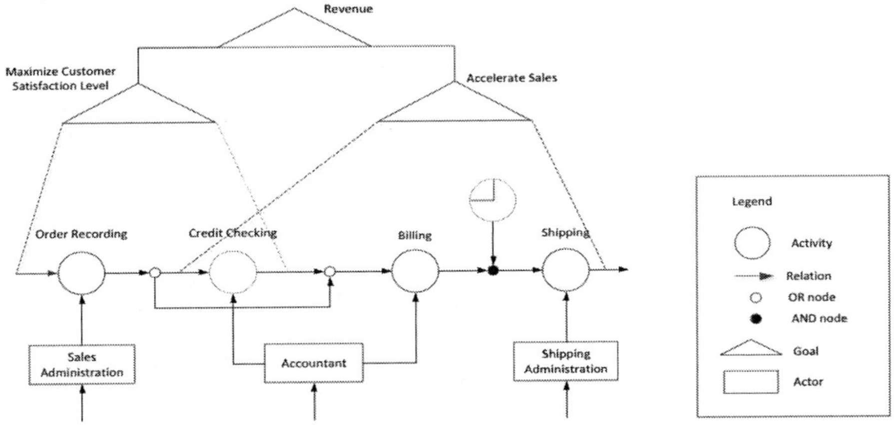

Source: Goal Based Workflow (Kavakli, 2003)

Figure 1 Goal Based Workflow

2.3 Keep All Objects Satisfied (KAOS)

KAOS might be described as the framework from several paradigms that enables to combine various different level of thought along with its reasoning. KAOS modeling language is part of KAOS framework to elicit, specify, and analyze goals, requirement, scenario and tasks responsibility [4].
The Element of KAOS [4] consists of following terms:

- Goal is defined as the set of behavior or circumstance that have to be fulfilled or accepted by the system in the predefined condition [7]. The goal definition must be clear enough such that could be verified if the system is able to fulfill that goal.
- *Softgoal* is used to documentions the system's alternative behavior, such that its satisfaction could be verified. The satisfaction level of the softgoal is restricted by the predefined limit.
- An agent is the object that has a function to process the activity operation. An agent, that is an active component, could represent human, hardware, software, that has specific role in achieving the goal.
- There are three kinds dependencies between goals in KAOS [4]:
- AND/OR-decomposition is the relation that describes the hierarchy of goal and its sub-goal. It explains a goal is achieved if all of its sub-goal is achieved (using AND decomposition) or if there is at least one of its sub goals is achieved (using OR decomposition).
- Potential conflict is the relation that express in certain condition the achievement of a goal would hinder the achievement of another goal. This conflict usually appears if there are any differences among stakeholder's interest and point of view [7].
- Responsibility assignment is the relation between agent and a goal. The agent has the responsibility to achieve the goal.

2.4 iStar

The iStar(i*) approach is centered on the stakeholder of the system and also the relationship between stakeholder. Through this approach the requirements of the systems could be learn easily, especially if there is specific requirement in the domain, and also preparing the change of requirement in the future [2].

iStar approach express the dependability between actor and the goal that should be achieved, task that should be done, and resource that should be provided. In addition of goal, softgoal is also introduced to express non-functional requirement. The i* name itself stand for "distributed intentionality, in which i* place the intention of the actor in the context of social network between actors [4]. Those actor could be referred to human or machine. Actors are described as entity that can act freely, restricted by its relation with another actor, cannot be known or controlled fully, and have explicit knowledge to exercise certain action [4]. In order to describe the actor entity, i* complements it with motivation and intention attribute [4]. Instead of focus on actor behavior, intention and relation property is used to describe the actor. Actor has a relation with another actor, due to there is goal that want to be achieved, task that should be done, and resource that should be provided.

Dependency that describe the intentional relationship between two actor are represented in following term:

Actor1→ Dependum → Actor2

Where "dependum" means kind of dependency and explain the deal between two entities. The first actor is the party that needs something from the second actor. Type of dependency can be categorize into 4 types [4]:

- Goal is the condition or circumstance to be achieved by the stakeholder. It is stated as emphasizing in representation language. How to goal could be achieved is not defined, therefore the alternative should be considered.
- Task defines the certain course to do some action. If task is defined as the subcomponent of higher task, it restricts the higher task to certain action.
- Resource is an entity (physically or information). The main concern here is the availability of the resource and identifies the provider if external party provides it.
- Softgoal is the condition that has subjective level of satisfaction. Its purpose is not defining precisely only follow its interpretation.

iStar Modeling consists of:

1) *Strategic Dependency* (SD), describes the dependency relationship between actor, that is the strategy of each actor to understand:
   a. What do they need?
   b. How they can achieve their needs?
   c. To whom they depend in achieving their need?

2) *Strategic Rationale* (SR), modelize what is happening inside actors, and how the actors achieve their need. SR models consists of:
   a. *Mean – end link*: represents the task that need to be done to achieve the final result. The final result could be goal, task, resource or softgoal.
   b. *Task decomposition link*: represents the relation between task and its subcomponent. Its subcomponent could be goal, task, resource and softgoal.

Contribution link: represents how positive is the goal's, resource's, or task's contribution to the achievement of certain softgoal.

2.5 RACI (Responsible, Accountable, Consulted, Informed)

RACI Chart is the matrix that describes the role of parties in completing certain work on a project of a business process. This matrix is very useful in explaining the role and responsibility of entity in a project or process. RACI itself is stood for Responsible, Accountable, Consulted and/or Informed. As can be seen in figure 2 RACI chart [8]:

RACI Chart

Activities \ Functions	CEO	CFO	Business Executive	CIO	Business Process Owner	Head Operations	Chief Architect	Head Development	Head IT Administration	PMO	Compliance, Audit, Risk and Security
Link business goals to IT goals.	C	I	A/R	R	C						
Identify critical dependencies and current performance.	C	C	R	A/R	C	C	C	C	C		C
Build an IT strategic plan.	A	C	C	R	I	C	C	C	C	I	C
Build IT tactical plans.	C	I		A	C	C	C	C	C	R	I
Analyse programme portfolios and manage project and service portfolios.	C	I	I	A	R	R	C	R	C	C	I

A **RACI** chart identifies who is **R**esponsible, **A**ccountable, **C**onsulted and/or **I**nformed.

Source: IT Governance, 2007

Figure 2 RACI chart

Following are the description of RACI chart value:

- Responsible: The person or entity who exercise the activity.
- Accountable: The person or entity that responsible has the authority to take the decision in certain circumstance.
- Consulted: The person or entity that his feedback and advice is required to complete the activity.
- Informed: The person or entity that has the right to know the result of the activity or the decision.

3 Research Methodology

The perspectives and experiences of the researcher influence a research design. These views and experiences are used to guide how and for whom the research problem is being investigated, what research approach and methods of gathering suitable research data will be used.

Author will focus on developing new modeling framework based on the existing goal and intentional based modeling framework. The comparison between existing modeling framework, and how each model influences the proposed modeling framework will be described. Finally, the author will make a modeling framework and describe the sequence process needed to implement the framework to the real case.

4 Role Based Goal Orientation

In this study, this model analyzed based on four criteria: description of goal, actor, softgoal, and task and event. Analysis of requirement engineering models in describing the goal, actor, softgoal, task and event notation and briefly summarized in Table 1 is a table of analysis among models.

From the summary Table 1, models will be chosen is the model that will become a basic reference in the design requirement engineering approach and must have the ability to represent a goal and derivatives of these goals in detail relations and cooperation between the actors in the domain that will be described. Model iStar is the right model to serve as a basic reference in this study. But iStar models also have shortcomings, i.e cannot represent the role of the actor, therefore needed a concept that can represent the role of the actor. Based on the results of literature study RACI diagram is an appropriate concept to be adopted in this modeling.

RACI in an organization is necessary, because it describes positions, roles and functions of each actor in the organization. RACI defines the responsibility of each actor in the organization for each business process. RACI allows the business process designer to allocate actor's responsibility on every business process vary from each other according to the needs of the business process, which makes RACI is flexible. RACI can also identify if there is overload of work on an actor, so the distribution of responsibility can be optimized.

Therefore it can be concluded that the concept of RACI is highly appropriate to fulfill deficiencies in this model because RACI is useful to identify actors and relationships between actors in the modeling that describes the dependence of the actors of the goal to be achieved, the task will be done and the resources that will be provided.

Table 1 Table of comparison among models

Model	*Goal*	Actor	*Softgoal*	*Task* and *Event*
KAOS, 1990	Has 4 models representation.In modeling goal, task is not expressed.	Only knows agent concept. Cooperation between agents can not be observed explicitly. There is not actor role in the process.	*Softgoal* is represented as goal. The constraint in achieving the goal is expressed as obstacle.	*Task* has more detail representation in operational model, by expressing the relationship between event and task.
***Goal Based Workflow*, 1994**	Has goal representation, which is derived into simpler goal.	Introduces actor concept, but the relationship between actor can not be observed. There is not actor role in each of process.	Does not introduce softgoal concept.	Task and event concept are represented by activity. The model expresses the sequence of activity in achieving the goal.
iStar, 1995	Has goal and softgoal representation, and also expresses task and another entities in the same model.	Shows the cooperation between actors explicitly. There is not further exploration on role and position of actor (even though there is available syntax)	Has softgoal representation, and negative or positive contribution relationship, by which the progress of softgoal achievement can be observed.	Model iStar can not express the operational side of view, which is the sequence of event in achieving the task.

4.1 Deconstruction Model

Translation of the model modification from models deconstruction can be seen in Table 2, Table Deconstruction Model

Table 2 Table Models Deconstruction

Model/Concept	Proses	Output
Goal Based Workflow	Deconstruction	• Sequence
KAOS	Deconstruction	• Task

Model/Concept	Proses	Output
iStar	Pra Deconstruction	• Strategic Rationale Notation • Strategic Dependency Notation
iStar: Strategic Rationale	Deconstruction	• Goal • Resource
iStar: Strategic Dependency	Deconstruction	• Actor
RACI	Adoption	• Role

From the results of Table 2 the obtained models proposed are composed by elements of the output of the process from deconstruction on every model.

To further improve the modeling is made, then the first deconstructed into elements essential constituent of the proposed model.

The proposed model consists of two notations models:

1. Strategic Dependency; In building the strategic dependency notation on the proposed model used the following considerations:
   a. The basic model used is a model of strategic dependency on istar.
   b. iStar strategic models doesn't show the dependency on the role of each actor on each process, requiring modifications to the notation to describe the role of the actor on every process.
   c. In the proposed model uses the concept of RACI diagram to explain the role of each actor, and then roles in the RACI diagram notation on the model adopted as proposed strategic dependency.
2. Strategic Rationale; In building the strategic rationale notation on the proposed model used the following considerations:
   a. The basic model used to create the strategic rationale notation is modeled on istar strategic rationale model.
   b. Strategic Rationale model on istar illustrates the task that will do by each actor in order to achieve the goal of every actor, but does not show the sequential order of the task performed by the actor.
   c. To illustrate the sequential task sequence then adopted elements of Goal-Based Workflow modeling. KAOS has event concept that is used as a substitute for the task on Strategic Rationale for on proposed model.

4.2 Design Process Model

The proposed model called Role-Based Goal Oriented Model is derived from the characteristics of the model created. Role-Based Goal Oriented Model has two main notations the same as the model iStar(i*), namely strategic dependency and strategic rationale that has been modified.

Strategic dependency consists of six stages, as depicted in figure 3. At the end of this stage, strategic dependency notation is obtained as the final result (see on Appendix A).

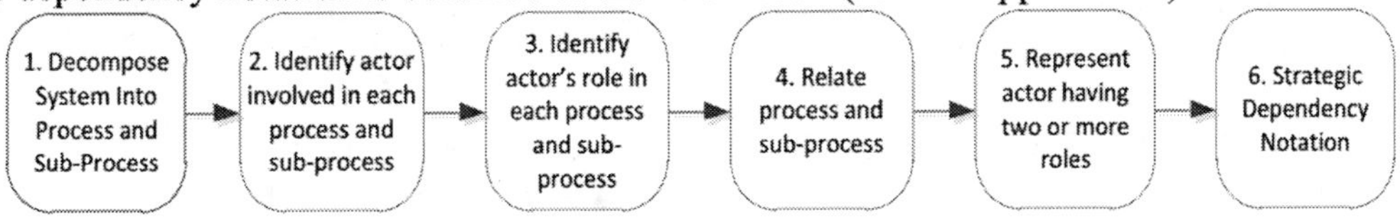

Figure 3. Stages Modelling Notation proposed Strategic Dependency

Furthermore, the strategic rationale modelling stage consists of four stages, as depicted in figure 4. At the end of this stage, strategic rationale notation is obtained as the final result (see on Appendix A).

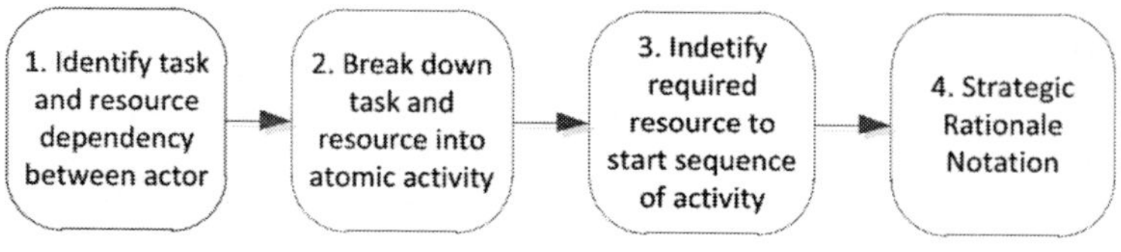

Figure 4. Stages Modeling Notation proposed Strategic Rationale

Having obtained a complete figure of the internal activities inside the actor, the next step is to draw notation strategic rationale. One strategic rationale notation represents one entry in the column actors. Sequences of activity become a task on the strategic rationale.

5 Conclusion and Further Research

5.1 Conclusion

Based on the discussion of the research that has been done, produced some of the conclusions that can be delivered as follows:

1. Modeling iStar at the time of requirement engineering can be more easily understood, especially aware of any special requirements in a domain and also can determine the understanding of the relationship between actors. But in the absence of modeling iStar furthers exploration of the role and position of actors in each process and cannot display the operational side, the series of events that occur in living a task.
2. By modifying modeling Role modeling Based iStar be Goal Oriented proven model can also determine the dependence between actors and also the role and position of actors in each process shown by strategic dependency and can identify what activities are happening at each actor sequentially so it is clearly visible construction sequences to achieve the goal in the internal processes of the actors represented by the strategic rationale.

5.2 Further Research

Suggestions for further research are as follows:

1. In Role Model Based Goal Oriented Construction needed a tool that is appropriate to adopt this model.
2. The development of this model can also be used in different organizational forms that involve actors who are part of the organization.

References

[1] Curtis, B., Krasner, H., & Iscoe, N. (1988). A Field Study of the Software Design Process for Large Systems. Communications of the ACM, 31(11), pp.1268-1287

[2] Yu, Eric, Paolo Giorgini, Neil Maiden, and John Mylopoulos. (2011) : Social Modelling for Requirement Engineering: An Introduction. *Social Modelling for Requirements Engineering*. MIT Press.

[3] Nuseibeh, Bashar, and Steve Easterbrook. (2000): Requirements Engineering: A Roadmap. *ICSE 2000 Proceedings of the Conference on The Future of Software Engineering*

[4] Yu, Eric: Social Modeling and i*. *Conceptual Modeling: Foundations and Applications*. Springer, 2009.

[5] Sandfreni, Oktadimi, N. R., Surendro, K. (2015): Requirement Engineering for Cloud Computing in University using i*(iStar) Hierarchy Method. Information Science and Application. Lecture Note in Electrical Engineering, pp 885-890.

[6] Kavakli, E., dan Loucopoulos, P. (2004). Goal Modelling in Requirements Engineering: Analysis and Critique. *Of Currents Methods, Informastion Modelling Methods and Methodologies.* 104-124

[7] Hui, B., Liaskos, S.m Mylopoulos, J., "Requirements Analysis for Customizable Software: A Goals-Skills-Preferences Framework". Proc. International Conference on Requirements Engineering (RE'03), Monterey, USA, September 2003.

[8] Janssen, M. & Cresswell, A. (2005) An Enterprise Application Integration Methodology for E-Government. Journal of Enterprise Information Management, 18(5), pp.531-547.

Appendix A

Strategic Dependency Notation

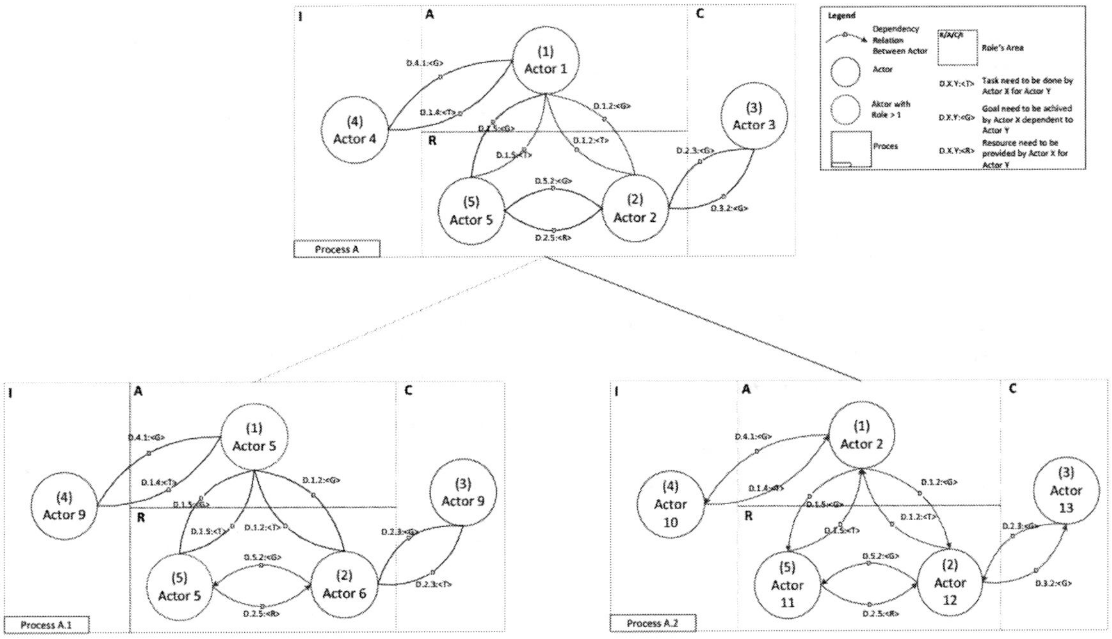

Strategic Rationale Notation

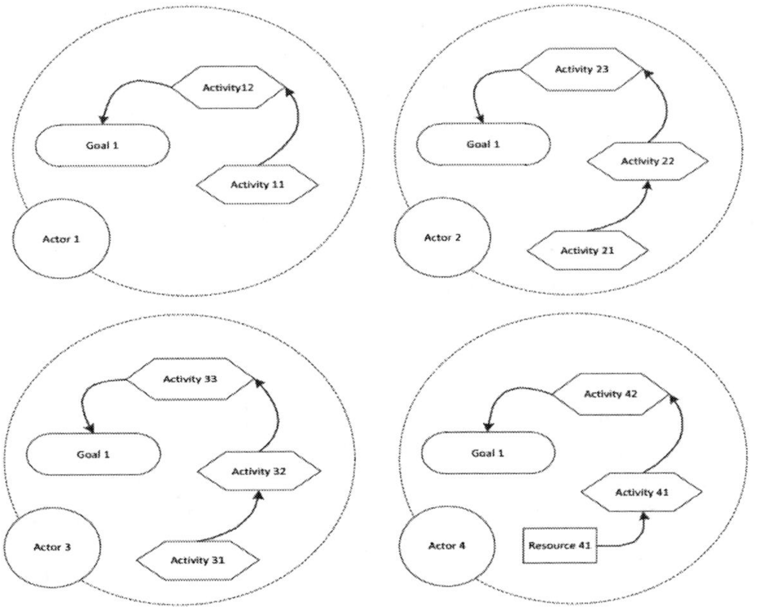

MATEC Web of Conferences **52**, 01005 (2016)
DOI: 10.1051/matecconf/20165201005

Ant Foraging Behavior for Job Shop Problem

Diyana Abdul Mahad[1] and Zuhaimy Ismail[2]

[1]*Quantitative Methods Unit, Faculty of Management, Multimedia University, 63100 Cyberjaya, Selangor Darul Ehsan, Malaysia*

[2]*Department of Mathematics, Faculty of Science, Universiti Teknologi Malaysia,81310 Skudai, Johor Darul Ta'zim, Malaysia*

Abstract. Ant Colony Optimization (ACO) is a new algorithm approach, inspired by the foraging behavior of real ants. It has frequently been applied to many optimization problems and one such problem is in solving the job shop problem (JSP). The JSP is a finite set of jobs processed on a finite set of machine where once a job initiates processing on a given machine, it must complete processing and uninterrupted. In solving the Job Shop Scheduling problem, the process is measure by the amount of time required in completing a job known as a makespan and minimizing the makespan is the main objective of this study. In this paper, we developed an ACO algorithm to minimize the makespan. A real set of problems from a metal company in Johor bahru, producing 20 parts with jobs involving the process of clinching, tapping and power press respectively. The result from this study shows that the proposed ACO heuristics managed to produce a god result in a short time.

1 Introduction

Scheduling has been defined as "the art of assigning resources to tasks in order to ensure the completion of these tasks in a reasonable amount of time" [1]. A job shop scheduling problem is a problem of as scheduling n given jobs varying sizes on m given identical machines, while trying to minimize the total length or time of the schedule (that is when all the assigned jobs have finished processing). In a finite job shop setting, it has a finite set of jobs on a finite set of machines. Each job is characterized by a fixed order of operations; each of the jobs is to be processed on a specific machine and specified duration. Each machine can process at most one job at a time [2]. Once a job initiates processing on a given machine it must complete processing and must be uninterrupted. It is common that nowadays, the job command are given online where upon completing a job, a system or an algorithm is required to make a decision about the next job. Basically, the Job Shop Problem (JSP) is divided into three common shop models for scheduling namely Open Shops, Flow Shops, and Job Shops. Unlike JSP and FSP, the Open Shop model is a 'build to order' system where no inventory is required.

JSP is a discrete problem of a combinatorial problem and has been widely studied. It is scheduling problem generalized from the Travelling Salesman Problem (TSP). It has also been illustrated as a problem in computational complexity theory which is hard to solve in practice and well known as an NP-hard problem. The general objective of scheduling is to decide in what rules, methods or order to perform jobs in the sequence, so that certain desired criteria are met. In this research, near-optimal solution methodologies for job shop scheduling are examined.

In searching for the optimal or near optimal solution to the JSP, we explore the possibility of using Ant Colony Optimization (ACO) approach. It is a new algorithm inspired by the foraging behavior of real ants. Many researchers have used and developed ACO algorithm to solve NP-hard problem such as travelling salesman problem [1], graph colouring [3], vehicle routing problem [4] and so on. The main aim of this study is to minimize the makespan that is the total completion time [5]. The makespan can be described as the time required to complete all the n jobs. Minimizing the makespan is a common objective in multiple-machine sequencing problems. A real scheduling problem from a small and medium enterprise producing components for auto industry is used. The rest of the paper is as follows. The full description of JSP is given in section 2 followed by the mathematical modeling. In Section 3, a brief explanation of the ACO approach, the background of the ACO and the pseudo code in the methodology flow chart. In Section 4, the discussion on the adaptation of ACO to solve JSP and the results and discussion is given in the final section.

2 Problem Dissertation

In JSP, it must have a finite set of n jobs, and finite set of m machines. Each job must have a chain of operations, and each machine can be handling on at most one operation at a time. Each operation needs to be processed during an uninterrupted period length of time on a given machine.

The problem of scheduling jobs in a machine shop is often modeled as a *classical* job shop problem, which is described as follows. Given is a shop consisting of m machines $M_1, M_2,\ldots, M_m$. On these machines, a set of n jobs $J_1, J_2\ldots J_n$needs to bescheduled. A schedule defines for each job J_j, a completion time C_i such that the jobs do not overlap in their execution [6]. Each machine is available from time 0 onwards and can process at mostone job at a time. Each job $J_j$consists a chain of operations $O_{1j}, O_{2j},\ldots,O_{nj}$, wheredenotes the number of operations of job J_j. Operation O_{ij}can only be processedafter the completion of operation O_i–1, j (i=2,...,n_j); operation $O_{1j}$is available fromtime 0 onwards. O_{ij}needs uninterrupted processing on machine m_{ij}during a givennon-negative time p_{ij}. The objective is usually to find a schedule that minimizes themakespan, that is, to find a schedule in which the time to process all jobs is minimal.

This paper refers to the standard model of the n-job, m-machine job shop problem as;

$$J / M / G / C_{max}$$

where J and M stand for the number of jobs and machines, and the G and C_{max} represent the precedence rules and the minimum makespan. Respectively C_{max} is the maximum earliest completion time of the last operation of any job.

An example of $3/3/G/C_{max}$ JSP with three jobs and three machines is shown in Figure 1 with the matrix description given in Table 1.

Table 1: The Matrix represents the description on figure above.

	M_1	M_2	M_3
J_1	12	5	9
J_2	2	9	10

Table 1 show the data of an instance with 3 machines and 2 jobs, where each job consists of 3 operations. Whereas Figure 1 show the representative of the Table 1 into the graph. From Figure 1 we can see that the graph have 8 nodes, 17 edges and 6 operations. This has been calculates from equation below. From Equation 1, 2, and 3 we know that n is stand for jobs, m is for machines, and $|O|$ is operations it takes.

$$\text{Nodes} = n * m + 2 \quad (1)$$

$$\text{Operation } |O| = n * m \quad (2)$$

$$\text{Edges} = \frac{|O| \cdot (|O| - 1)}{2} + n \quad (3)$$

Figure 1 shows the example of the 2 jobs processed with the 3 machines. As the results, the crossing path for the problem is 0- 1 -2 – 6 - 7, and the makespan, as the total completion time is 20.

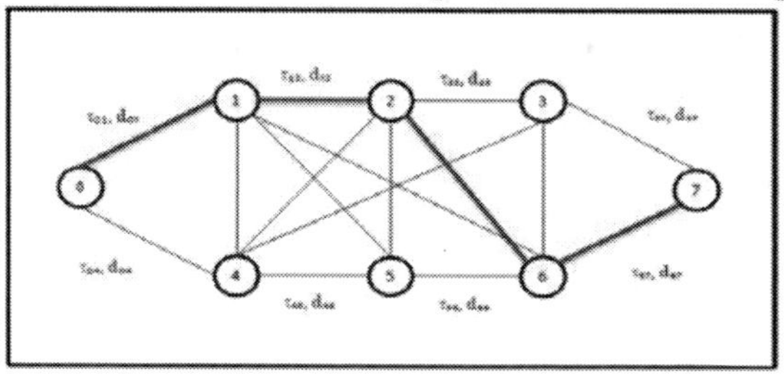

Figure 1. The graphic representation for the result of a $2/3/G/C_{max}$job shop problem.

2.1 The Mathematical Model

The mathematical formulation for the job shop with the objective of minimizing makespan are presented below:
Objective:

$$\text{Minimize}\{\text{Max}\ (C_{1j}, C_{2j}, C_{3j}, \dots, C_{ij})\} \quad (4)$$

Subject to:

$$C_{ij} - S_{ij} - t_{ij} = 0 \qquad \forall(i; j) \quad (5)$$

$$C_{i'j'} - C_{ij} + H(1 - \gamma_{ij}) \geq ti'j',\ \forall(i, j), (i', j') : Oij \in N_k,\ Oi'j' \in N_k \quad (6)$$

$$C_{ij} - C_{i'j'} + K(1 - \gamma_{ij}) \geq tij,\ \forall(i, j), (i', j') : Oij \in N_k,\ Oi'j' \in N_k \quad (7)$$

$$S_{ij}, \geq 0\ \forall i; j \quad (8)$$

$$Sij{+}1 - Cij \geq 0,\ \forall i, j = 1, \dots, Ji - 1 \quad (9)$$

$$\gamma = \begin{cases} 1, & \text{if operation} O_{ij}\ \text{precedes} O_{i'j'} \\ & 0,\ \text{Otherwise} \end{cases} \quad (10)$$

The above model detail the assumptions for the JSP. In Equation 4, let S_{ij}be as the starting point, while $C_{ij}$is the completion time job i. While H and K is assumed as very large positive integer, $N_k$is the set of operations O_{ij}that can be loaded on machine k and $\gamma$is a decision variable that generates a sequence betweenthe operations Oij and $Oi'j'$. The constraint sets in Equation 3, imposes that thedifference between the completion time and the starting time of an operation is equalto its processing time. This constraint satisfies the assumption that once an operationhas started, it cannot be pre-empted until its completion. Constraint sets in Equation 4and Equation 5 are to ensure that no two operations can be processed simultaneouslyon the same machine. The disjunctive constraint in equation 4, becomes inactivewhen $\gamma = 0$ and the disjunctive constraint in Equation 5 becomes inactive when$\gamma = 1$. Constraint set in Equation 6, are to ensures that the start time of an operationis always positive. Constraint set in Equation 7 represents the precedence relationshipamong various operations of a job.

Some Properties are :

Property 1: An optimal schedule exists if no idle time between any consecutive jobs for scheduling jobs on a single machine with common due dates.

Property 2: For the optimal schedule, V-shaped property exists around the common due date. This means that jobs completed before or on the common due date d are scheduled in non-increasing order of the ratios λ_i/α_i, and jobs starting on or after d are scheduled in non-decreasing order of the ratios λ_i/α_i in an optimal schedule.

Property 3: In an optimal schedule, either the first job starts at time zero or the completion time of one job coincides with the common due date d.

3 The Methodology

The ant colony optimization was inspired by a colony of real ants in an experiment conducted by Goss *et al.* A laboratory colony is given access to a food source in an arena linked to the colony's nest by a bridge with two branches of different length (see Figure 2).

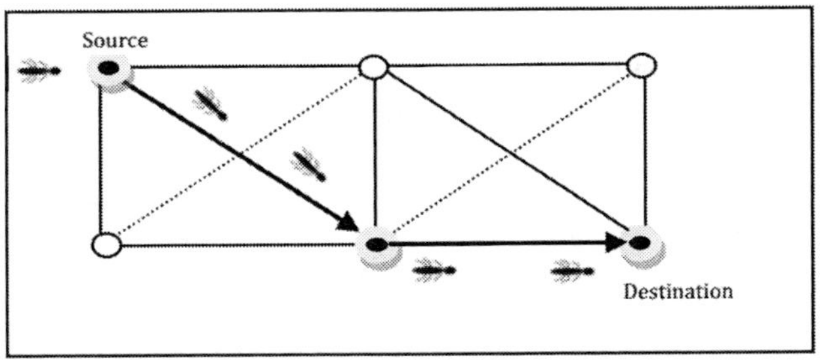

Figure 2. Experiment for the bridge experiment.

Figure 2, also shows how the ants move to on each arc *(i,j)* of the graph, that is associated with a variable τ_{ij} called *artificial pheromone trail.* Pheromone trails are read and written by ants. The amount (intensity) of pheromone trial is proportional to the utility, as estimated by the number of ants using that arc to builds good solutions. Each ant applies a step-by-step constructive decision policy to build problem's solution. At each node local information, maintained on the node itself and/or on its outgoing arcs, is used in a stochastic way to decide the node to move on. The decision rule of an ant k located in node i uses the pheromone trials τ_{ij} to compute the probability with which it should choose node $j \in N_i$ as the next node to move to, where N_i is the set of one-step neighbors of node i:

$$p_{ij}^{k} = \begin{cases} \tau_{ij} & \text{if } j \in N_i \\ 0 & \text{if } j \notin N_i \end{cases} \tag{11}$$

While building a solution ants deposit pheromone information on the arcs they use. In ACO, ants deposit a constant amount $\Delta\tau$ of pheromone. Consider an ant that at time t moves from node i to node j. It will change the pheromone value τ_{ij} as follows:

$$\tau_{ij}(t) \leftarrow \tau_{ij}(t) + \Delta\tau \tag{12}$$

Using this rule, which simulates real ants' pheromone depositing on arc (I, j), an ant using the arc-connecting node i to node j increases the probability that ants will use the same arc in the future. As in the case of real ants, autocatalysis and different path length are at work to favor the emergence of short paths.We have to write the data in a more compact form to tackle this job shop problem.

4 ACO Implementation

In this flowchart, we are adding two dummy nodes which classify the start and finish node; as the starting node and the completion of the overall job shop. Thus, now we have the starting point and the destination in the ACO. This flowchart is define as the Ant Colony algorithm. At the beginning of the algorithm, all parameters are initialized. There are m ants in the colonies that are placed at the starting points, and one iteration is completed when all ants has arrived at the destination. The local pheromone update when an ant chooses an edge to go on. The global pheromone updating is performed when the whole ant colony reached the destination. Those step influences the next iteration of ant colonies finding routes. The algorithm stopped when the maximum number of iterations reached, and the minimum makespan of the job shop is found.

In implementing the model developed, this study explores a real problem of JSP in a Metal Company. This Metal Company involved in the covers product design and development, electrical and electronic enclosure as well as industrial equipment for workshop, warehouse and factory. This company was a premier supplier for the metal components.Data given the general information for 20 parts of Customised E07 Rack which is commonly used to store computers for industrial production.

Initialization: The pheromone trails, the heuristic information and the parameters are initialized
Iterative loop:

2.1 A colony of ants determines starting jobs.
2.2 Construct a complete schedule for each ant:
Repeat
Apply state transition rule to select the next processing job
Apply the local updating rule
Until a complete schedule is constructed
2.3 Apply local search process
2.4 Apply the global updating rule

Cycle: If the maximum number of iterations is realized, then STOP;
Else go to step 2.

5. Results and Analysis

The objective of our study is to determine the minimum makespan, using the ACO. Based on the ACO method, a computerised system was develop and built using the Visual Basic programming language. Several trial run on the system shows that it is very efficient for small size problem but it reduces the average as the problem sizes involves. Table 2 shows the different process, jobs, makespan and the sequence in the JSP.

Table 2. The Results for Makespan Objective for Customized E07 rack.

Process	Total of Jobs	Makespan(seconds)	Machine	Job Sequence
CNC Machining	15	600	1	1, 7, 13
			2	2, 8, 14
			3	3, 9, 15
			4	4, 10
			5	5, 11
			6	6, 12
Bending	13	104	1	1, 7, 13
			2	2, 8
			3	3, 9
			4	4, 10
			5	5, 11
			6	6, 12
Clinching	3	72	1	1
			2	2
			6	3
Tapping	2	40	3	1
			5	2
Power Press	5	25	2	5
			3	4
			4	3
			5	2
			6	1

For instance, we analyse all the results that receive. We do through each the results on minimizing the makespan. Hence, there are several ways to estimate the running time of a program. To simplify the analysis, we will adopt the convention that there are no particular units of time. Thus, we throw away leading constants. We will also throw away low-order terms, so what we are essentially doing is computing a long running time. Since this is the long running time, we must be careful, never to underestimate the running time of the program. In fact, the answer provided is a guarantee that the program will terminate within a certain time period. The program may stop earlier than this, but never later[7].

Since, we had the heuristic method for the algorithm, there was the parameters that should be aware. We should do some parameter adjustment for theα, β, and ρthat we put in the algorithm. The underlying principle of this heuristic method is that a job with higher total processing time should be given higher priority than a job with less total processing time [8].We should increasing and decreasing the parameters, rather than just to follow the mathematical formulation can also be considered in order to make dynamic so that it is more powerful and can be applied to any range of set of data. Lastly, this implement of the program are very friendly user. However, the best way of balancing these features differed from problem to problem [9]. This program can be use, even there are different case study, which we discuss in this study. Future research for other case study should be done and compiling some improvement of the program.

References

1. M. Dorigo, L.M. Gambardella (1997). *Ant colony system: A cooperative learning approach to the traveling salesman problem, IEEE Transactions on Evolutionary Computation 1,* page 53–66.
2. Chang, R.-S., Chang, J.-S. and Lin, P.-S. An ant algorithm for balanced job scheduling in grids. *Future Generation Computer Systems*, 2009. URL http://www.elsevier.com/locate/fgcs.
3. E. Salari, K. Eshghi (2005). *An ACO algorithm for graph coloring problem, in: Congress on Computational Intelligence Methods and Applications,* page 15–17.
4. Xiaoxia Zhang, Lixin Tang (2005) *CT-ACO hybridizing ant colony optimization with cycle transfer search for the vehicle routing problem, in: Congress on Computational Intelligence Methods and Applications*, page 15–17.
5. WIKIPEDIA, Website "http://en.wikipedia.org/wiki/Job_shop" Retrieved on 12 September 2008.
6. J.A. Hoogeveen, S.L. Van De Velde, (1991): *Scheduling Around a Small Common Due Date. European Journal and Operational Research 55*, page 237–242.
7. Z. Ismail, D. Abd Mahad (2009). *Ant Colony Optimization for minimizing makespan for job shop problem.* Simposium Kebangsaan Sains ke-17. Melaka.
8. Betul Yagmahan, Mehmet Mutlu Yenisey. (2008). *Ant Colony Optimization forMulti-objective Flow Shop Scheduling Problem.* Turkey.
9. Kathryn A. and Dowslanda. (2008). *An improved ant colony optimisation heuristic for graph coloring.*

MATEC Web of Conferences **52**, 02001 (2016)
DOI: 10.1051/matecconf/20165202001

Comparative Calibration of Corrosion Measurements Using K-Nearest Neighbour Based Techniques

Yaman Hamed[1,a], A'fza Shafie[1],Zahiraniza Bt Mustaffa[2]and Naila Rusma Binti Idris[3]

[1]*Fundamental and Applied Sciences Department, Universiti Teknologi PETRONAS, Tronoh, Malaysia*
[2]*Civil Engineering Department,Universiti Teknologi PETRONAS, Tronoh, Malaysia*
[3]*Upstream Technical Services (UTS), Technlogy & Engineering Division,Kuala Lumpur, Malaysia*

Abstract. Every measuring equipment or inspection tool is known to have its own accuracy, which may affect the reliability of its measurements. This includes oil and gas pipeline corrosion defects measurements. The inspection tolerance occurred in the measurements should be treated carefully for each equipment to prevent misinterpretation of the data which could lead to incorrect assessment. This paper presents a comparison between two K-Nearest Neighbour (KNN) interpolation techniques used to calibrate corrosion measurements collected by Magnetic Flux Leakage Intelligent Pig (MFL-IP) with the readings of Ultrasonic Testing (UT) scan device. The comparison has relied on the position of the interpolators, the weight sequence, and the error in the final enhanced metrics compared to the original measurements. Both techniques have the potential to calibrate and enhance IP measurements, with relative advantage for one technique in reducing over fitting problem. This enhancement will be used to improve the integrity assessment report that depends on the disturbed corrosion metrics of oil and gas pipelines, to decide whether the pipeline is fit for service or needs certain maintenance.

1 Introduction

The early detection of an operational oil pipeline defects can prevent failure which can lead to huge catastrophes. Corrosion is one of the major factors that can lead to a pipeline system failure; hence it is important to understand the actual condition of the operational pipeline via proper inspection and using the correct tools. Non-destructive corrosion detection techniques vary according to the purpose of the inspection, accessibility to the pipeline system, level of accuracy and the reliability of data required[1].

Smart/intelligent pigs are used to provide information about the conditions of a pipeline and can be used to locate the problem areas. Metal-loss inspection pigs are used to detect defects that have resulted in wall thinning in the pipeline. There are two main types of metal-loss pigs i.e. magnetic-flux leakage (MFL) and ultrasonic testing (UT). (MFL-IP)when compared to other scan techniques, is widely used as an in-line Inspection (ILI) in the corrosion detection field considering it's applicability in both offshore and onshore pipelines and its capability to discriminate defects to some extent, given its comparative cost and time consumption. UT scan technique is normally applied externally and is used to inspect localized section. UT is known to have better sizing accuracy since absolute

[a] Corresponding author : yaman_85@hotmail.com

measurement can be obtained when compared to MFL which provides relative measurements[2].Table 1 shows the sizing accuracy for different corrosion detection techniques [3].

Table 1. Sizing accuracy of MFL and UT tools

Inspection tool	Sizing accuracy	Confidence level
HR-MFL IP	± 10% T	80 %
XHR-MFL IP	± 5 % T	80 %
HR-UT Scan	± 6 % T	80 %
XHR-UT Scan	± 3 % T	80 %

HR: high resolution, XHR: extra high resolution, T: wall thickness.

As shown in Table 1, both MFL and UT technologies have certain limitations regarding the accurate measurement of the defects within the system since both devices suffer from a certain *error margin* depending on the device's sizing accuracy. Since UT measurements have a smaller sizing accuracy than IP measurements, a consideration of the UT device being more accurate than MFL-IP can be done. This measuring error affects the integrity assessment and may lead to an over or an under-estimation of the actual condition of the scanned pipeline. In this paper, by using two different techniques, K-nearest neighbor interpolation method (KNN) was used to enhance the inaccurate readings of the corrosion metrics collected by the MFL-IP device, with the metrics collected by a UT device, bearing the consideration that MFL-IP devices suffer from a wider error margin than the one affecting UT devices. A comparison between the two methods was applied to show which interpolator achieves better representation of the original measurements towards the goal ones.

2 Background on interpolation

Interpolation is the process of predicting a missing or an unknown value of a function or a sample point using the known points around it (neighbors)[4]. Different techniques can be applied as interpolators; Polynomial interpolation, Multivariate interpolation, Bilinear interpolation, Bi-cubic spline interpolation, K-Nearest-neighbor interpolation (KNN), Inverse distance weighting (IDW) Interpolation, quadratic interpolation, B-spline interpolation, Lagrange interpolation, Gaussian interpolation, among other techniques[5-7]. Interpolation techniques are well expressed in the application of image processing, data mining, artificial neural networks, as well as other variant applications. This paper, presents a new usage of interpolation techniques as calibration of the measurements of two non-destructive corrosion inspection tools. More precisely, K-Nearest Neighbor interpolation (KNN) was used to enhance the accuracy of corrosion readings collected by MFL-IP tool using the measurements of a UT scan device.

Nearest neighbor method is a statistical test that is used to determine the significance of a point's nearest neighbor in order to calculate the deviation from the general trend. Nearest neighbor algorithm selects the nearest point value and does not take into account the values of other neighboring points, hence producing a constant interpolation[5, 8]. Nearest neighbor technique is based on a comparison between the distribution of distances between a studied point and its nearest neighboring points in a set of randomly distributed data. The Distance function used in the comparison is a function that defines a Euclidean distance between each pair of elements of a set. The Euclidean distance between two points, $x = (x_1, \dots, x_n)$ and $y = (y_1, \dots, y_n)$ can be seen in (1):

$$d=\|x-y\| = \sqrt{\sum_{i=1}^{n} (x_i - y_i)^2} \tag{1}$$

where$d$ is the distance function, n is the sample size [9, 10].

The predicted point using the nearest neighbor method is simply the value of the nearest point among the n points of a sample to the unknown one, which is:

$$\hat{y}=y_i \;\; ; \;\; d_i=\{min \; \|\hat{y}-y_i\| \;\; ; \; i=1,2,\dots n\} \tag{2}$$

The use of only the nearest point to predict the missing one, will exclude the effect of the other neighbor points which may lead to a biased estimator, herein to include the effect of multi neighbors to the predicted point and in order to reduce the effect of outliers of a sample point, K-nearest neighbor method is suitable for this purpose.

K-Nearest neighbor method: this method differs from the aforementioned nearest neighbor technique by considering the contiguity of the surrounding points to the required point. The contiguity can be estimated using what is called the weight function, which is defined as a function that measures the effect of each one of the neighbor points on the required one. In other words the estimated value of the missing or required point is the weighted average of its neighbors[11]. Loftsgaarden and Quesenberry introduced the KNN weight function in the relative field of density estimation[12], followed by Cover and Heart for the purpose of classification. The simplest weight function can be described as the ratio of the distance between each point of the neighborhood to the total sum of distances[11]:

$$w_i=\frac{d_i}{\sum d_i} \quad ; \quad d_i=\|\hat{y}-y_i\| \tag{3}$$

Franke & Nielson [13], used the classical form of weight function as seen in (4):

$$w_i=\frac{h_i^{-p}}{\sum h_i^{-p}} \tag{4}$$

where$p$ is a positive real number chosen randomly and is called the power parameter (usually equals to 2), and h is the distance from point i to the interpolated point from the original data set.

Shepard [5], had used a modified weight function for superior results compared to the original one, as shown in(5):

$$w_i=\frac{\left[\frac{R-h_i}{Rh_i}\right]^2}{\sum\left[\frac{R-h_i}{Rh_i}\right]^2} \tag{5}$$

where$h_i$is the distance from the interpolated point to point i of a set of random data. R can be defined as the distance from the interpolated point to the farthest point of the set of data, and n is the number of the neighbors.

Hinton & Roweis [14], introduced a Stochastic Neighborhood embedding algorithm, an algorithm that uses a weight for the nearest neighboring points as seen in(6):

$$w_i=\frac{e^{-d_i}}{\sum e^{-d_i}} \tag{6}$$

whered_i is the distance between the interpolated point and its i'th neighbor.

Jianping Gou et al. [15], used a combination of the ratio between the interpolated point and its furthest neighbor, to the distance of its nearest neighbor as in (7):

$$w_i=\frac{d_i-d_{max}}{d_i-d_{min}} \times \frac{d_i+d_{max}}{d_i+d_{min}} \tag{7}$$

whered_i is the distance between the interpolated point and the point i, d_{max} is the distance between the interpolated point and furthest point in the set, d_{min} is the distance between the interpolated point and the nearest point in the set.

The interpolated point using the K-Nearest Neighbor algorithm can be defined by (8):

$$\hat{y}_i=\sum w_i y_i \tag{8}$$

wherew_i is the proper weight for each one of the neighbor points y_i to the interpolated point $\hat{y}_i$ [11].

The methods described in this paper were built on Hinton's & Roweis's weight whilst using KNN. The exponential function defined in (6) will maximize the effect of the nearest neighbor to the point that requires calibration, and by that, the interpolator will get the properties of the closest neighbor point, compared to the other surrounding points.

3 Proposed method

A 20 year operational oil pipeline used in Malaysia with a diameter of 25.4 cm, length of 3.9 kilometers, and internal wall thickness of 12.7 mm, was chosen to collect its internal corrosion measurements. Data on the corrosion geometry parameters (depth, length, width) were collected using two In-Line Inspection (ILI) tools; MFL-IP with a sizing accuracy of ±20% of the corrosion inspected measurements, and UT scan device with a sizing accuracy of ±5%.

Corrosion measurements collected from two different devices have the same properties however differ in formation. The measurements collected by a UT scan device is represented in a grid of the remaining wall thickness along the pipeline length, (the grid is set at constant dimensions horizontally and vertically) as shown in Table 2.

Table 2. Tabulation of UT corrosion measurements

			10:43	10:39	10:35	10:31	10:27
	Log Distance	X (mm)	0	5	10	15	20
1	108.705	-290	11.36	11.36	11.45	11.26	11.24
2	108.710	-285	11.36	11.36	11.31	11.24	11.22
3	108.715	-280	11.31	11.31	11.24	11.17	11.17
4	108.720	-275	11.19	11.19	11.24	11.29	10.94
5	108.725	-270	11.26	11.26	11.24	11.45	11.08
6	108.730	-265	11.4	11.4	11.24	11.36	11.19
7	108.735	-260	11.26	11.26	11.29	11.33	11.17

Table 3, shows the measurements collected by MFL-IP, which are represented as single points with relative metal loss, position and orientation.

Table 3. Tabulation of MFL-IP corrosion measurements

s-log distance [m]	s-o'clock position [h:m]	type	dimension	depth [%]	depth [mm] 11.1	length [mm]	width [mm]	ERF	at int. pipewall	Remaining wall thickness
108.527	0.320	MELO	PITT	11.000	1.221	15.000	15.000	0.819	YES	11.48
108.577	0.242	MELO	PITT	28.000	3.108	15.000	15.000	0.822	YES	9.59
108.582	0.195	MELO	PITT	13.000	1.443	20.000	20.000	0.821	YES	11.26
108.607	0.210	MELO	PINH	14.000	1.554	10.000	10.000	0.818	YES	11.15
108.637	0.359	MELO	PINH	10.000	1.110	10.000	10.000	0.818	YES	11.59
108.662	0.320	MELO	PITT	10.000	1.110	15.000	15.000	0.819	YES	11.59

To start the calibration process, each point was joined with its adjacent to form a pair, this is done by applying a mapping procedure to both of the corrosion measurement sets. The mapping procedure started from the same pipeline log-point for each device,which was recorded in the measurements report for each defect. The distance between the defect and the device's Log-point is illustrated in Fig.1.

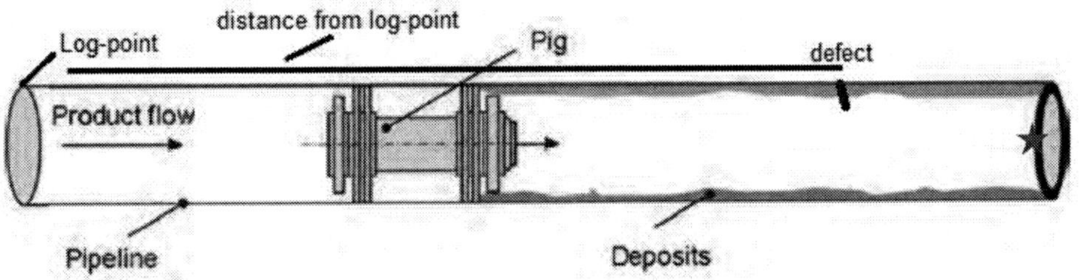

Figure 1. Distance from log-point to defect

This is followed by detecting the exact position of each pair by using a clockwise orientation method within the pipeline as shown in Fig. 2.

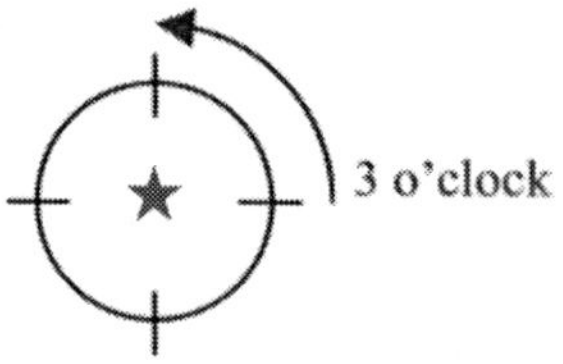

Figure 2. Clockwise orientation

Since corrosion measurements that are collected byMFL-IP are given as relative measurements compared to the UT device which are represented as a grid of actual remaining pipeline wall thickness, and in the light of the large sizing accuracy that UT device has compared to MFL-IP, the measurements of the UT device were assumed to be the goal measurements used to calibrate the measurements of the MFL-IP. In this paper bothproposed methods took in consideration a grid of the closest 8 measurements to the UT point to represent the neighbors of the *actual/goal* corrosion measurement. Fig.3 shows the distribution of points with respective neighbors where point ut_5 (the 5th element) is equivalent to is the goal for the point that requires calibration collected by MFL-IP.

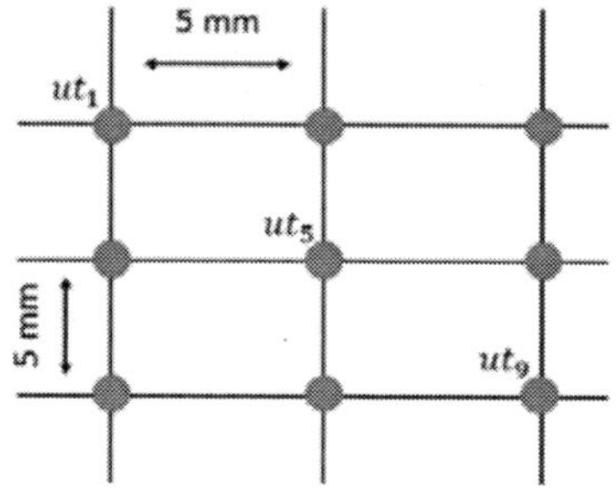

Figure 3. UT neighbor's grid

The first KNN method used in this paper to calibrate corrosion metrics of MFL-IP considered the center of UT measurements neighborhood as the position of the point that required the calibration. Using the Exponential weight function described by Hinton & Roweis, with the Euclidean function to describe the distance between the neighbors, (EwEf) method was applied as follows:

A. The neighbors grid were expanded by squeezing MFL-IP measurements in each grid as the 6'th element in the matrix, the point is placed such that it may seem to replace the original, however in actuality it is at a distance very near to the 5th point hence the vector now consists of 10 elements instead of 9, and the neighbors became 9 instead of 8 as shown in Fig. 4 and Fig. 5,hence will be used as $k$the number of neighbors in KNN interpolation. If $\{ y_i ; i: 1 \dots 10 \}$ is the variable representing the expanded remaining wall thickness matrix, then y_5 and y_6 represent the UT goal point and the IP point that requires calibration respectively.

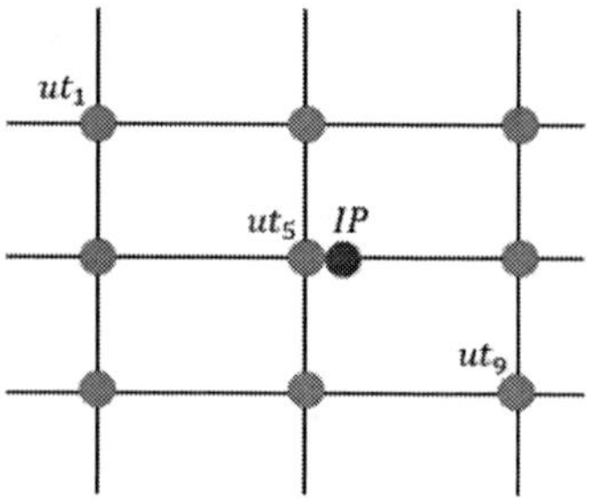

Figure 4. Expanded grid with IP measurements

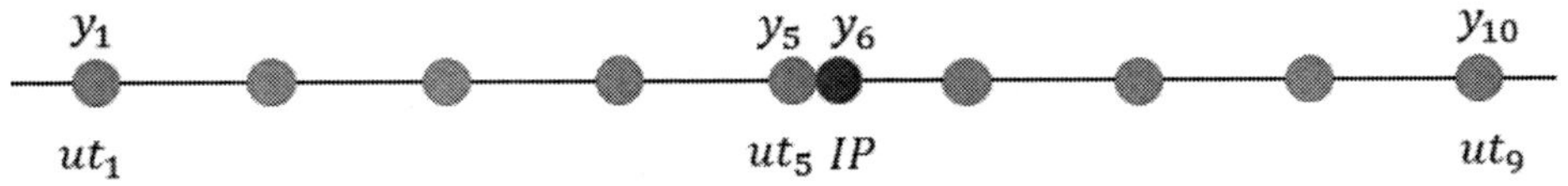

Figure 5. The vector representing remaining wall thickness expanded matrix

B. The distance between the center point (ut_5) and its neighbors were calculated using the Euclidean distance function as in (1).
C. A weight sequence was calculated for the expanded vector by using Hinton & Roweis weight function (6). The exponential function included within (6) will concentrate on the center point as it has the biggest effect on the predicted point. Adding to that, since the center point is the goal point, the trend well be stronger towards the actual required measurements.
D. By applying KNN interpolation technique (with k=9) on the 5th element in the vector, equation (8), will replace the IP value with the weighted average calculated by KNN technique which will result in an output that is the new calibrated corrosion measurements closer to the actual measurements collected by the UT scan device.
E. A comparison between the original UT measurements, original MFL-IP measurements, and the new interpolated corrosion measurements was conducted to determine the amount of enhancement that was done with the suggested method.

As mentioned earlier, EwEf calibration method is a method that applies KNN interpolation technique with an assumption that the position of the center point in the neighborhood (ut_5), is the position of the point that requires calibration, since (ut_5) is the goal point. However this assumption may lead to a biased estimator of the calibrated point towards the center of the grid most of the time, which could result in data over fitting. In light of this concern, the second method used in this paper to calibrate MFL-IP measurements used different position for the point that requires calibration in the UT neighborhood grid. The method was applied by assuming that the position of the point that requires calibration (IP) should consider the relationship between the original IP and UT measurements to avoid the over fitting possibility. To achieve that, a Maximum-Minimum based calibration method with Exponential weight and Euclidean distance functions (MMEwEf), was applied as follows:

A. if the MFL-IP measurement was larger than UT measurement, then the position of the point that require calibration will be assumed to be the position of the Maximum point in the neighborhood. On the contrary, if MFL-IP measurement was smaller than UT measurement, then the position of the point that require calibration will be assumed to be the position of the Minimum point in the neighborhood. The movement of the calibrated point's position will assure that the enhanced point after calibration will be relatively closer to the original MFL-IP measurement rather than the goal measurement (ut_5), which will reduce the possibility of a biased estimator. Fig 6, and Fig 7, show the distribution of the point that require calibration with UT measurements within the neighborhood grid in both cases respectively.

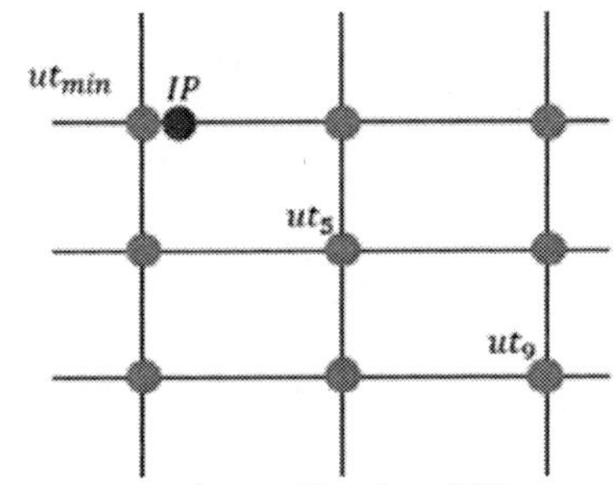

Figure 6. Neighbors' grid when IP measurement is smaller than UT.

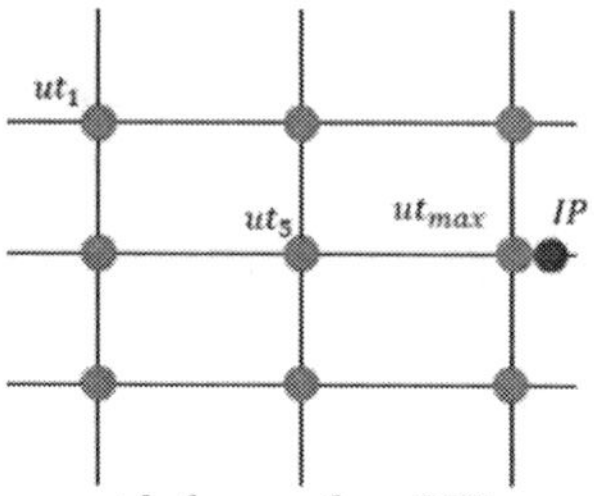

Figure 7. Neighbors' grid when IP measurement is larger than UT.

B. The expanded grid was used similarly (as with EwEf method) to generate a weight sequence with Hinton & Roweis weight function described in (6), and Euclidean distance function as in (1).

C. The weight sequence then was used to calibrate (IP) measurements by (8), the data set resulted by applying (8) is considered the calibrated corrosion measurements.

A comparison between the original UT measurements, original MFL-IP measurements, and the new interpolated corrosion measurements was conducted to determine the amount of enhancement that was done with the suggested method.

4 Results and discussion

The suggested methodology was applied on one segment of the studied pipeline, which contains 31 defects as reported in the reliability assessment report given by the MFL-IP operator. The calibration techniques used in this paper showed a remarkable enhancement in corrosion measurements. Fig. 8, and Fig.9 show comparisons between the original measurements, and the enhanced measurements using both proposed methods, EwEf, and MMEwEf calibration respectively. The enhanced IP measurements showed closer behavior to the goal UT points.

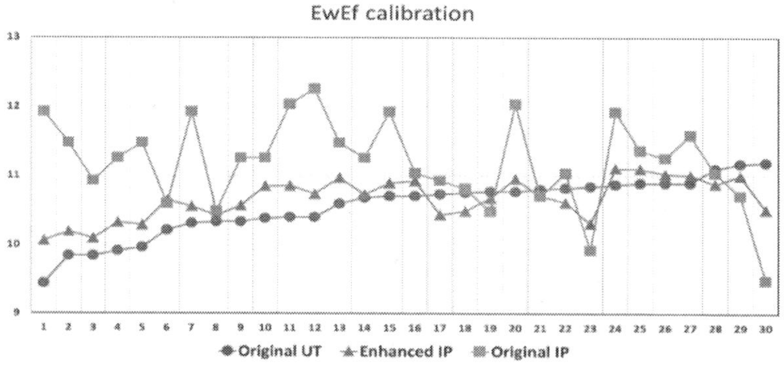

Figure 8. Corrosion measurements calibration by EwEf method

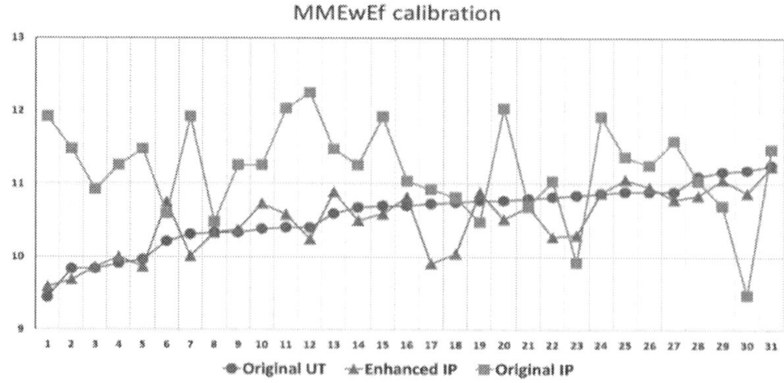

Figure 9. Corrosion measurements calibration by MMEwEf method

Table 4 shows the minimum remaining wall thickness in the studied segment of the pipeline collected by the UT device (actual size), compared to the thickness recorded by theMFL-IP device at the same point, and the thickness after the enhancement with both suggested methods.

Table 4. Comparison between minimum remaining wall thickness as recorded by different devices

	Minimum wall thickness		
Method	MFL-IP	Actual UT measurement	Calibrated measurement
EwEf	11.923 mm	9.44 mm	10.06 mm
MMEwEf	11.923 mm	9.44 mm	9.58 mm

The part of the pipeline that has the thinnest wall thickness is considered as the weakest point in the whole system. The point that has a minimum wall thickness of an operational pipeline is usually used in the extreme value analysis, which is applied to determine the probability of system failure. Table 4 shows the minimum wall thickness recorded using the Ultrasonic scan device as 9.44 mm. while the measurement reported by MFL-IP for the same point was 11.923. The calibrated measurement for the same point using the proposed EwEf, and MMEwEf calibration methods was found to be 10.06 mm, and 9.58 which is 14.66%, and 18.44% better than the original metrics respectively.

This difference between the calibrated measurement of the point that has the minimum wall thickness in the segment, and the original IP corrosion size using EwEf method is found to be 1.86 mm, which means that the service life of the studied segment is 4 years shorter than it was reported by the MFL-IP, considering that the average corrosion growth rate per year regarding to NACE (National Association of Corrosion Engineers) is reported to be 0.4 mm [16]. In other words, using the reported MFL-IP measurement without calibration will give a statement that the segment is fit-to-serve for an extra time approximated to be 4 years, while the actual size give much shorter service time. This huge difference may be lead to a possible system failure considering the lack of information related to the sizing accuracy of the MFL-IP device.

Likewise, the difference between the calibrated, and the original IP measurement using MMEwEf method was found to be 2.34 mm, which means an even shorter service life of the studied segment, which could be estimated to be 6 years.

Table 5 shows the error in the calibrated measurements compared to the original MFL-IP points using the proposed techniques.

Table 5. Error in calibrated measurements.

	Calibration method	
Source of error	EwEf	MMEwEf
Calibrated IP to original IP	0.98 mm	0.84 mm

Since the error in the calibrated measurements using MMEwEf method is smaller than the error committed using EwEf method,this means that MMEwEf calibration showed less biased estimation towards the goal measurements compared to the EwEf method. The 0.84 mm is equivalent to 6.61% of the pipeline wall thickness, while 0.98 mm is equivalent to 7.71%. The difference in the error between the two techniques refers to the superiority of the MMEwEf method on EwEf method when it comes to the biased estimation. This reduction of the error in calibrated measurements is due to the position movement of the point that requires calibration in the neighbors grid. The interpolator got the properties of the closest neighbor point to its own properties which reduce the error caused by KNN interpolation. This error reduction could be described as reducing the over fitting possibility in the calibrated measurements.

5 Conclusions

KNN interpolation technique can be used as proposed in this paper to calibrate corrosion measurements collected by different devices. The methods proposed in this paper showed remarkable results in reducing the error in corrosion measurements collected by MFL-IP compared to the actual corrosion metrics collected by UT scan device. Both EwEf, and MMEwEf calibration methods presented in this paper showed that using KNN will enhance the original IP corrosion measurements by 14.66%, and 18.44%, with an overall error of 6.61% and 7.71% of the actual pipeline wall thickness respectively. While EwEf calibration estimate closer measurements to the goal data, MMEwEf method reduces the bias in the estimators by considering the relationship between the original measurements. Further sample points testing should provide wider understanding to the differences between the proposed methods, since both had showed close results in the field of measurement calibration.

References

1. A. Barbian and M. Beller,*18th World Conference on NondestructiveTesting*, Durban, South Africa, (2012)
2. A. Pople and P. Wharf, *Pipeline Rehabilitation and Maintenance*, (2003)
3. P. Ivanov, Z. Zhang, C. Yeoh, L. Udpa, Y. Sun, S. Udpa, et al,*IEEE Transactions on Magnetics*,**34**(1998)
4. P. Miklos, *2nd Siberian-Hungarian Joint Symposium On Intelligent Systems*, (2004)
5. P. D. Dumitru, M. Plopeanu, and D. Badea, *1st European Conference of Geodesy & Geomatics Engineering GENG'13*, Antalya, Turkey,(2013)
6. R. Olivier and C. Hanqiang, "Nearest neighbor value interpolation," arXiv preprint arXiv:1211.1768, (2012)
7. T. M. Lehmann, C. Gönner, and K. Spitzer, IEEE Trans. Med. Imag, **18**,(1999)
8. R. Franke, Math. Comp, **38**, (1982)
9. K. Q. Weinberger, J. Blitzer, and L. K. Saul, *Advances in neural information processing systems*, (2005)
10. J. Walters-Williams and Y. Li, *Advanced Techniques in Computing Sciences and Software Engineering*, (2010)
11. W. Härdle and O. Linton, *Handbook of econometrics*, **4**, (1994)
12. D. O. Loftsgaarden and C. P. Quesenberry, Ann. Math. Stat, **36**,(1965)
13. R. Franke and G. Nielson, INT. J. NUMER. METH. ENG, **15**, (1980)
14. G. E. Hinton and S. T. Roweis, *Advances in neural information processing systems*, (2002)
15. J. Gou, L. Du, Y. Zhang, and T. Xiong, J. Inf. Comput. Sci,**9**, (2012)
16. F. Caleyo, A. Valor, V. Venegas, J. H. Espina Hernandez, J. C. Velazquez, and J. M. HALLEN, OIL GAS J, **110**, (2012)

MATEC Web of Conferences **52**, 02002 (2016)
DOI: 10.1051/matecconf/20165202002

A Study on a Multi-Objective Optimization Method Based on Neuro-Response Surface Method (NRSM)

Jae-Chul Lee[1], Ji-Ho Jeong[1], Hicham Kharoufi[1] and Sung-Chul Shin[1,a]

[1] *Department of Naval Architecture and Ocean Engineering, Pusan National University, Jangjeon 2-dong, Geumjeong-gu, Busan, South Korea, 609-717*

Abstract. The geometry of systems including the marine engineering problems needs to be optimized in the initial design stage. However, the performance analysis using commercial code is generally time-consuming. To solve this problem, many engineers perform the optimization process using the response surface method (RSM) to predict the system performance, but RSM presents some prediction errors for nonlinear systems. The major objective of this research is to establish an optimal design framework. The framework is composed of three parts: definition of geometry, generation of response surface, and optimization process. To reduce the time for performance analysis and minimize the prediction errors, the response surface is generated using the artificial neural network (ANN) which is considered as NRSM. The optimization process is done for the generated response surface by non-dominated sorting genetic algorithm-II (NSGA-II). Through case study of a derrick structure, we have confirmed the proposed framework applicability. In the future, we will try to apply the constructed framework to multi-objective optimization problems.

1 Introduction

The optimal design of engineering system is considered as multi-objective optimization problem, and the geometry of engineering system severely affects their performance. For this reason, determining an optimal geometry is one of the challenging problems in the initial design stage. In recent years, system optimization based on the performance using the commercial code is a method that was employed for engineering design problem [1, 2].

The essence of optimization design process is the performance analysis/evaluation according to the geometry modification. Generally, the performance analysis of the complex engineering system such as marine, aerospace and machinery systems is time-consuming. To reduce the performance calculation time, many researchers try to predict the performance using approximation models. These approximation models represent the relationship between inputs and outputs. The RSM is used to predict the system performance [3, 4, 5], but the method produces some errors in highly nonlinear problems [6, 7, 8].

The optimal design problems of marine system based on its performance involve highly nonlinear components, such as structural, hydrodynamic and vibration performances. Many researchers tried to optimize the hull form using the parametric design method. Zhang et al. [9] researched hull form

[a] Corresponding author : scshin@pusan.ac.kr

design using the parametric approach. They employed the non-uniform rational b-spline (NURBS) method to represent the hull form. Gregory et al. [10] tried to optimise the hull form in calm and rough water based on the parametric design method. Lee and Choi [11] tried to optimise the hull form in terms of hydrodynamic performance using the form parametric design method. Park et al. [12] tried to optimise hull form of KSUEZMAX using optimization techniques. The parametric design method for geometry representation is a useful technique for geometry optimization based on performance.

Up to now, many relevant papers about optimization based on system performance have been published. To predict the system performance, Shin [13] employed the neuro-fuzzy algorithm to predict the wake distribution, Xu [14] predicted the maritime safety using the artificial neural network (ANN), and Lee [15] tried the prediction for added resistance in waves using the genetic programming (GP). To optimize the shape based on performance, Kim [16] developed a framework to optimize the stern form based on CFD. In the shape optimization research, CFD analysis was just a means of checking performance of a design process, and it is very time-consuming. Therefore, optimization becomes difficult/impossible using CFD tools because we need to check the system performances for a large number of alternative design cases. To examine alternative design cases and minimise the performance analysis time, a multi-objective optimal design framework that includes performance prediction and an optimization process is essential.

To create an appropriate approximation model for highly nonlinear engineering system, Lee [17] compares the prediction accuracy of the response surface generated by RSM, the kriging method and ANN method, and the analysis of prediction result shows that the prediction values given by ANN are more accurate than those of the other models. We subsequently propose an optimal design framework comprising two principal phases:

1st Phase: To predict the system performance, we generate the response surface using ANN that is considered as NRSM in the proposed framework.
2nd Phase: We optimize the shape in the generated response surface using NSGA- II.

By means of a case study of constraint optimization problem, the effectiveness of the proposed optimal design methodology is verified in view of structural performances.

2 Framework for optimum design based on NRSM

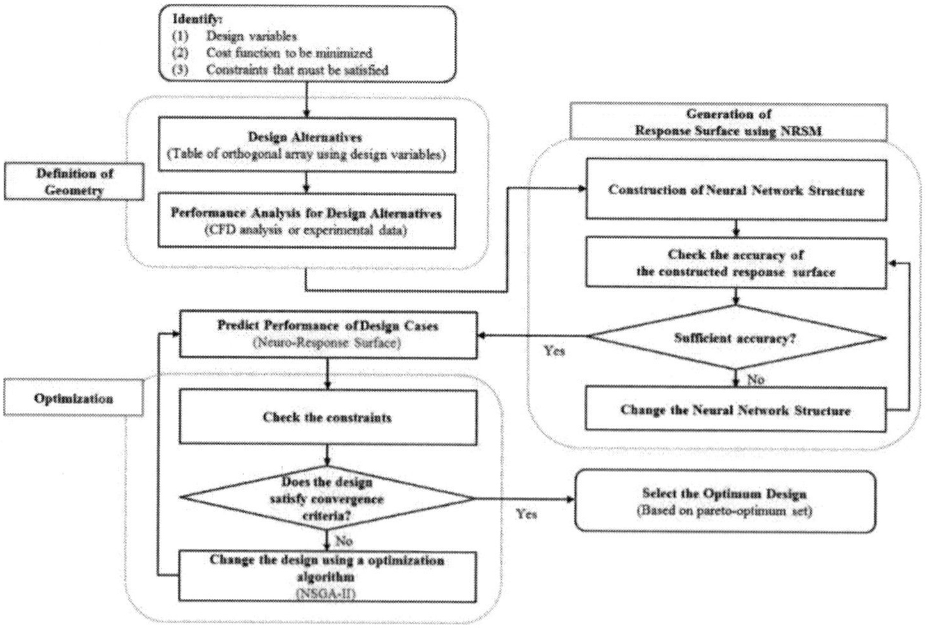

Figure 1. Optimal design framework based on NRSM [17]

The proposed optimal design framework (Fig. 1) includes three processes:

(1st process: Definition of the geometry)

The framework defines the geometry of the structure using parameterisation. An orthogonal array table [18] is used to systematically generate design alternatives. To check the interaction between the variables and the response, the generated design alternatives are divided into two sets:

- Training data: used to generate the response surface and check the learning accuracy.
- Test data: used to check the prediction accuracy.

(2nd process: Generation of response surface using NRSM)

The multi-layer perceptron (MLP) is used to generate the response surface. It has three layers: an input layer, a hidden layer, and an output layer. The back-propagation algorithm [19] is used to train the neural network. The prediction accuracy of generated response surface is very important because the optimization process is conducted on it. In order to construct the appropriate response surface, the best structure and the best number of learning cycles for the neural network are prepared and the prediction accuracy of the generated response surface is checked using Test data *(1st process)*. Through this process, we can make appropriate ANN structure for our problem.

After generating the response surface, the performance of various design alternatives can be predicted easily and quickly.

(3rd process: Optimization)

The optimization process is conducted to generate the response surface. The NSGA-II [20] is used as a multi-objective optimization algorithm.

Finally, the optimum structural design can be selected using the Pareto-optimum set generated by the proposed framework. The framework was constructed using MATLAB.

3 Case study

A derrick structure is used for a case study by considering its weight reduction while keeping its safety factor upper than 3. The accuracy of the framework results has been analysed using ANSYS.

3.1 Formulation of optimization problem

The equations below show the formulation process of the optimization problem (Eqs. 1~4):

Find x_i

x_i = Design variables (i = 1,2,3,4,5,6,7)

to minimize

$f_1(x)$ = Steel weight (ton)

to maximize

$f_2(x)$ = Safety factor

Using weighting factor, we can formulate the problem as one function to be minimized:

$$F(x) = (W_1 * f_1) + (W_2 * 1/f_2) \tag{1}$$

where,
W_i = Weighting factor (i = 1,2)

$$\textit{Subject to } \text{Min } x_i \le x_i \le \max x_i \text{ (i = number of design variables)} \quad (2)$$
$$f_1(x) \le \text{base design case} \quad (3)$$
$$f_2(x) \ge 3.0 \quad (4)$$

Derrick structure consists of 2 beam shapes (I beam and Box beam). Fig. 2 shows the cross section area shape and main dimension.

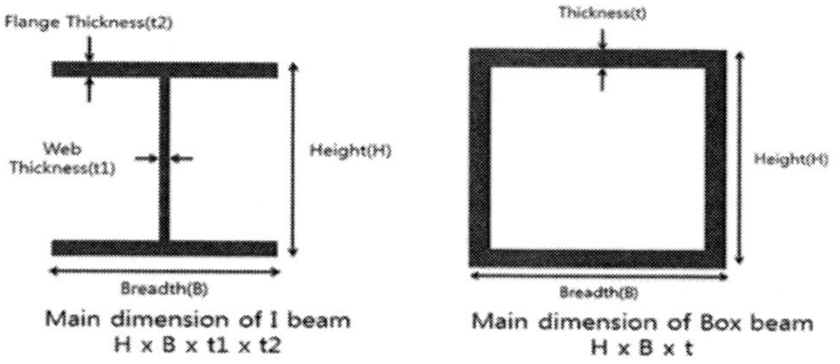

Figure 2. Cross section area shape and main dimension of I beam and Box beam

Seven design variables are considered (Table 1). Fig. 3 ((a) and (b)) shows the design model including the design variables.

Table 1. Design variables.

Breadth 750mm (I beam)	**Web thickness**	W1
	Flange thickness	F1
Breadth 500mm (I beam)	**Web thickness**	W2
	Flange thickness	F2
Top Box beam thickness		T
Middle Box beam thickness		M
Bottom Box beam Thickness		B

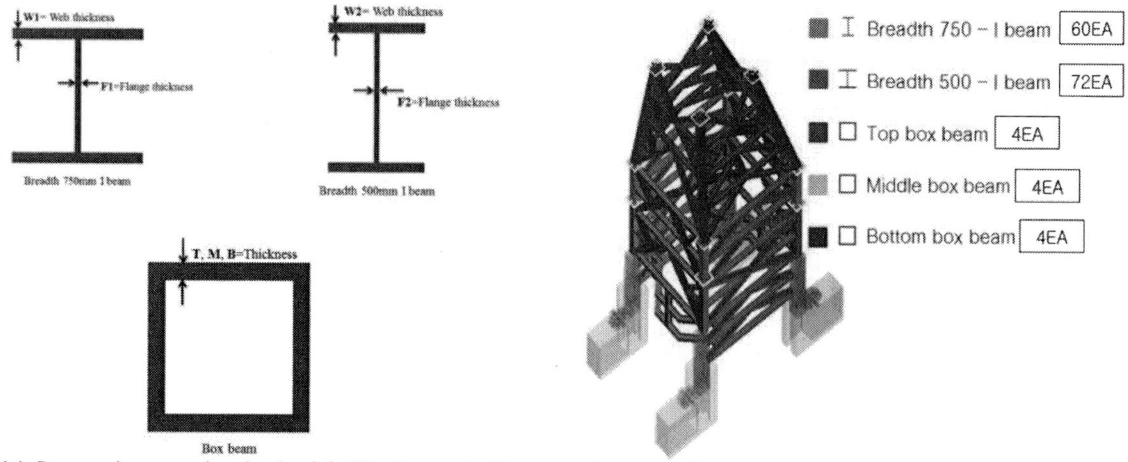

(a) Supporting member in derrick (Design model)

(b) Main dimensions for member (Design variables)

Figure 3. Design model and design variables

3.2 Environmental conditions

Fig. 4 shows the environmental conditions and constraints for calculation. A, B, C, D, E mean the gravity, acceleration of heave motion, rotational velocity of roll and pitch motion, vertical loads (wind pressure) and fixed condition respectively. Fig. 5 and Table 2 show the loading condition.

Table 2. Loading condition.

	HC, TDM and etc.	Pipe and etc.	Total
Weight	1,030ton	600ton	1,630ton
Mark (Fig.12)	A, B, C, D	E, F, G	-

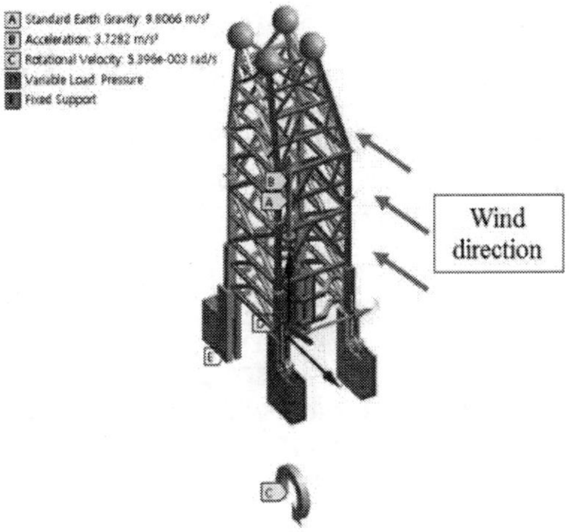

Figure 4. Environmental conditions and constraints

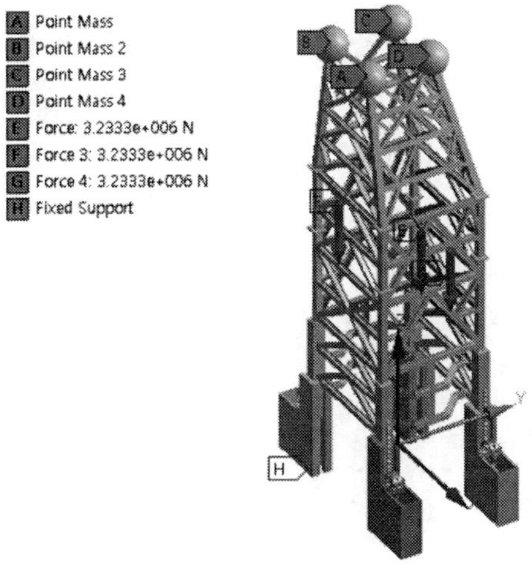

Figure 5. Loading condition

3.3 Definition of geometry

The proposed framework defines the shape of the structure by parameterization. The generated geometries are used to generate the response surface. 18 sets of different design alternatives are generated using an orthogonal array table (L_{18} ($2^1 \times 3^2$)). Table 3 shows design variables and the results of the performance analysis for steel weight and safety factor using the ANSYS. The safety factor is calculated as below:

$$\text{Safety factor} = \text{Yield stress} / \text{Maximum stress} \tag{5}$$

In equation (5), yield stress is a material property while the maximum stress is calculated on the derrick system.

Design no. 19 is considered as base design case.

Table 3. Design variables and performance analysis results.

Case	Design variables							Performance analysis results	
	W1	F1	W2	F2	T	M	B	Steel weight (ton)	Safety factor
1	25	40	20	30	80	80	80	1279.337	2.249
2	25	50	25	40	90	90	90	1439.202	2.746
3	25	60	30	50	100	100	100	1597.343	3.203
4	30	40	20	40	90	100	100	1390.810	2.431
5	30	50	25	50	100	80	80	1478.803	2.784
6	30	60	30	30	80	90	90	1491.352	3.126

7	35	40	25	30	100	90	100	1398.747	2.306
8	35	50	30	40	80	100	80	1460.352	2.728
9	35	60	20	50	90	80	90	1560.115	3.136
10	25	40	30	50	90	90	80	1393.590	2.390
11	25	50	20	30	100	100	90	1423.310	2.756
12	25	60	25	40	80	80	100	1498.982	3.195
13	30	40	25	50	80	100	90	1401.650	2.502
14	30	50	30	30	90	80	100	1440.041	2.717
15	30	60	20	40	100	90	80	1526.360	3.203
16	35	40	30	40	100	80	90	1420.795	2.328
17	35	50	20	50	80	90	100	1486.953	2.823
18	35	60	25	30	90	100	80	1511.466	3.174
19	35	60	30	50	100	100	100	1630.613	3.536

After performance calculation of the generated design cases, the response surface was generated using NRSM. Then, we predicted the performance of the design cases in a continuous response surface that was not directly computed. We used 15 sets of data to generate the response surface *(Case 1, 2, 3, 5, 6, 7, 9, 10, 11, 13, 14, 15, 17, 18 and 19)* and 4 sets of data to check its accuracy *(Case 4, 8, 12 and 16)*. To increase the learning rate for a neural network, all data was used as a normalized value between 0.5 and 1.

3.4 Generation of response surface using NRSM

The number of hidden layers was changed from 1 to 10. Using six hidden layers gave a better result. Therefore, the final structure and the number of learning cycles are 7-6-2 and 6400 respectively. Fig. 6 shows the error convergence in the learning process of the neural network. The error convergence (0.0004) occurs at approximately 6238 iterations.

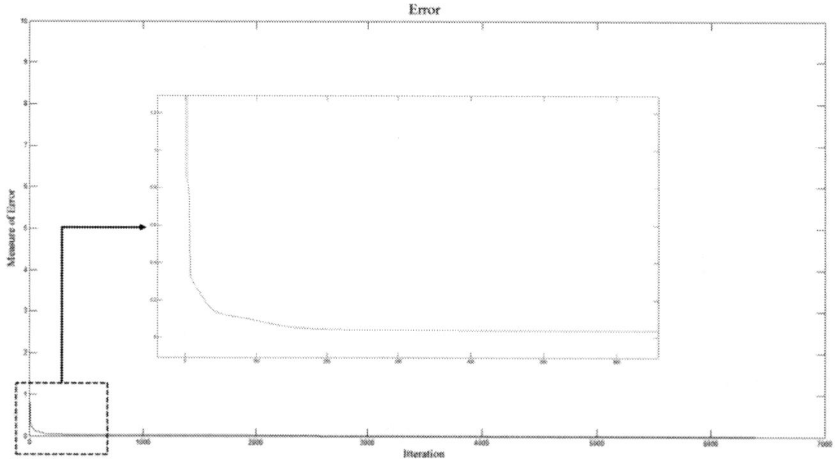

Figure 6. Measure of error

The error is defined in equations (6) and (7). The error value between the output of the network (d_j) and the actual value (y_j) is used. L is the number of output neurons.

$$e_i(n) = d_j(n) - y_j(n) \quad (6)$$

$$E(n) = \tfrac{1}{2} \sum_{j=1}^{L} e_j^2(n) \quad (7)$$

Table 4 shows the accuracy of the generated response surface using Training data set.

Table 4. Analysis results for training data set.

Case	Desired values (ANSYS)		Prediction values (NRSM)	
	Steel weight	Safety factor	Steel weight	Safety factor
1	0.500	1.000	0.500	0.992
2	0.728	0.752	0.720	0.746
3	0.953	0.591	0.962	0.587
5	0.784	0.736	0.782	0.728
6	0.802	0.615	0.799	0.607
7	0.670	0.966	0.665	0.965
9	0.900	0.612	0.893	0.612
10	0.663	0.919	0.656	0.919
11	0.705	0.747	0.698	0.751
13	0.674	0.861	0.666	0.868
14	0.729	0.764	0.724	0.771
15	0.852	0.591	0.850	0.601
17	0.796	0.721	0.797	0.724
18	0.830	0.600	0.833	0.604
19	1.000	0.500	0.979	0.501

The structure of the neural network is appropriate, because all of the error values *(Error = [(Desired value - Prediction values) / Desired values] ×100)* are below 2.1% *(Case 19)*. Table 5 and Table 6 show the prediction accuracy of the generated response surface and errors for the test data. Analysis of the results in Table 6 shows that there are still prediction errors. However, to determine the performance in a limited time, the NRSM can give reasonable results for the initial design stage.

Table 5. Analysis results for test data set.

Case	Desired values (ANSYS)		Prediction values (NRSM)	
	Steel weight	Safety factor	Steel weight	Safety factor
4	0.659	0.897	0.658	0.924
8	0.758	0.759	0.751	0.721
12	0.813	0.593	0.806	0.616
16	0.701	0.953	0.688	0.930

Table 6. Error of test data set.

Case	Error (%) = [(Desired value - Prediction values) / Desired values] ×100	
	Steel weight (ton)	Safety factor
4	0.2%	3.0%
8	0.8%	5.0%
12	0.8%	3.8%
16	1.9%	2.4%

3.5 Optimization

Table 7 and Fig. 7 show the parameters for NSGA-II and the pareto-optimum sets as the final result of the framework. To select the final optimum design among the pareto-optimum set, we used a weighting factor. 0.5 is the weighting factors of both Steel weight and Safety factor. The black point is the selected optimum design (Fig. 7).

Table 7. NSGA-II Parameter.

Parameter	Value
Population size	100
Generation	1000
Crossover	30%
Mutation	2%

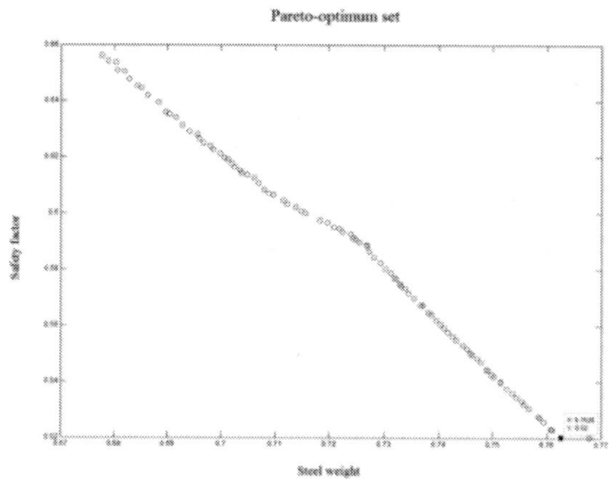

Figure 7. Pareto-optimum set

3.6 Analysis of Optimum Design

Table 8 shows the performance analysis results of the constructed framework and ANSYS result. When using the constructed framework, the prediction error is about 10% (Safety factor), as shown in Table 9.

Table 8. Result analysis.

Design variables for optimum design case						
W1 (mm)	W1 (mm)	W1 (mm)	W1 (mm)	W1 (mm)	W1 (mm)	W1 (mm)
25.000	25.000	25.000	25.000	25.000	25.000	25.000

NRSM Framework		ANSYS Calculation	
Steel weight (ton)	Safety factor	Steel weight (ton)	Safety factor
1463.848	3.457	1472.944	3.114

Table 9. Prediction error.

Prediction Error (%) = [(ANSYS Calculation - NRSM Framework) / ANSYS Calculation] ×100	
Steel weight (ton)	Safety factor
0.6%	10%

Fig. 8 shows the result of structure analysis. The maximum stress is 1.1238e2 MPa.

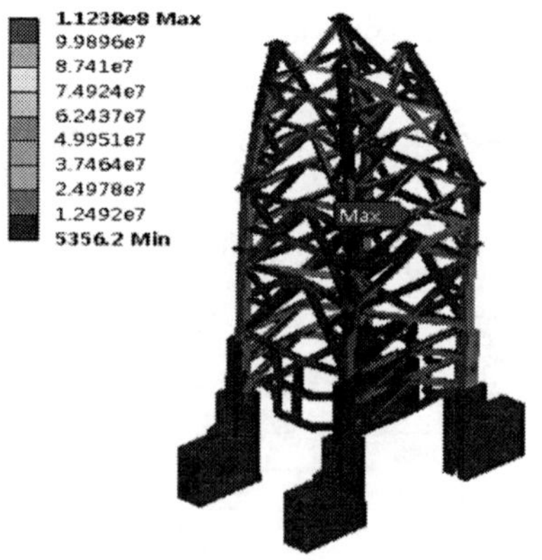

Figure 8. Structure analysis of optimum design

Finally, the improvement in standards of performance evaluation was analysed, as shown in Table 10, where all criteria for the optimum design case are satisfied such as steel weight that decreased about 10% comparing with base design case and safety factor that increased about 3.8% comparing with the required safety factor.

Table 10. Improvement of performance evaluation criteria.

Improvement (%) = [(Base model – Optimization model) / Base model] ×100	
Steel weight (ton)	**Safety factor (upper than 3.0)**
10% (Decrease)	3.8% (Increase)

3.7 Discussion

The effectiveness of the proposed framework through multi-objective optimization problem was confirmed by a derrick weight minimization while considering a given safety factor: more than 3.

In the initial design stage, the shape of various design alternatives considering their performances can be checked in a reduced analysis time using the proposed framework. We considered that optimal design framework presents the starting design point and save the time for optimal design using existing experimental or calculation data based on commercial codes in the initial design stage. However, the proposed framework shows some errors for extrapolation design problems. To increase the prediction accuracy of extrapolation design alternatives, we can add some experimental or calculation data using commercial codes into pre-existing data, or set the range of design variables including all possible design alternatives in the proposed framework *(1st process)*.

4 Conclusion

The following conclusions can be drawn from this study:

I. We proposed an optimization framework based on the NRSM:
   - The method generates the response surface using the ANN that is considered as NRSM in the proposed framework.
   - The system is optimised using the NSGA-II.

II. Through case study on the optimal structure of a derrick, we confirmed the feasibility of the proposed framework for multi-objective optimization problems:
   - Selected optimal design case satisfies the design constraints.

III. The proposed framework is considered as a useful engineering system optimization design tool in the initial design stage:
   - Save the performance analysis time in the optimization process;
   - Widely check the alternative design cases.

Evaluations of the appropriate approximation model for each problem and their application to various optimization problems considering actual design constraints will be conducted in future work.

Acknowledgement

This work was supported by the National Research Foundation of Korea (NRF) grant funded by the Korea government (MEST) through GCRC-SOP (No. 2011-0030671).

References

1. D. H. Ko, D. C. Ko, H. J. Lim, J. M. Lee, and B. M. Kim, I. J. P. E. M, **14** (8): pp 1301-1309, (2013)
2. S. P. Jung, Y. G. Kim, and T. W. Park, I. J. P. E. M, **13** (1): pp 57-63, (2012)

3. K.J. Hong, K. K. Jeon, Y. S. Cho, D. H. Choi, and S. J. Lee (2000), Trans. Korea Soc. Mech. Eng. A, **24**(6): 1408-1418, (2000)
4. R. H. Mayers and D. C. Montgomery, *'Response Surface Methodology – Process and Product Optimization Using Designed Experiments'*, John Wiley & Sons, (1995)
5. Z. Z. Li, T. H. Cheng, D. J. Xuan, M. Ren, G.Y. Shen, and Y. D. Shen, I. J. P. E. M., **13**(9):1641-1645, (2012)
6. M. Sankaya and A. Gullu, Journal of Cleaner Production, **65**: 604-616, (2014)
7. J.M. Salman, Arabian Journal of Chemistry, **7**(1): 101-106, (2014)
8. F. Kahraman, *Materials and Technology*, **43**(5): 267-270, (2009)
9. P. Zhang, Journal of Hydrodynamics, **20**(6) : pp. 804-810, (2008)
10. G. J. Grigoropoulos, D. S. Chalkias, Computer-Aided Design, **42** : pp. 977-984, (2010)
11. Y.S. Lee, Y.B. Choi, J. N. A. O. E., **46**(6): 562-568, (2009)
12. J. H. Park, J. E. Choi and H. H. Chun, Int. J. Nav. Archit. Ocean Eng, **7** : pp. 100-114, (2015)
13. S.C. Shin, Int. J. F. L. I. S., **17**(2) : pp.154-159, (2007)
14. J. H. Xu, *Information and Management Engineering Communications in Computer and Information Science*, **236**: 34-38, (2011)
15. J. C. Lee, J. H. Jeong, and S. C. Shin, Proceedings of the Annual Autumn Conference, S. N. A. K. : pp.482-490, (2014)
16. H.J. Kim, H.H. Chun, and H.J. Choi, J. S. N. A. K., **44**(6): 564-571, (2007)
17. J. C. LEE, *'Application of Multi-objective Optimization for Marine Systems using NRSM'*, Ph. D. thesis, Pusan National University, Busan, S. Korea, (2015)
18. P. J. Ross, *'Taguchi Techniques for Quality Engineering'* 2nd, McGraw-Hill, (1996)
19. H.N. Robert, IJCNN, **1**: pp. 593 -605, (1989)
20. K. Deb, IEEE TRAN. E. C., **6** (2): pp. 182-197, (2002)

MATEC Web of Conferences **52**, 02003 (2016)
DOI: 10.1051/matecconf/20165202003

Design Optimization of ESD (Emergency ShutDown) System for Offshore Process Based on Reliability Analysis

Jeong-hoon Bae[1], Sung-chul Shin[1,a], Byeong-cheol Park[1] and Soo-young Kim[1]

[1]*Department of Naval Architecture and Ocean Engineering, Pusan National University, Busan, South Korea*

Abstract. Hydrocarbon leaks have a major accident potential and it could give significant damages to human, property and environment.To prevent these risks from the leak in design aspects, installation of ESD system is representative. Because the ESD system should be operated properly at any time, It needs high reliability and much cost. To make ESD system with high reliability and reasonable cost, it is a need to find specific design method.In this study, we proposed the multi-objective design optimization method and performed the optimization of the ESD system for 1st separation system to satisfy high reliability and cost-effective.'NSGA-II (Non-dominated Sorting Genetic Algorithm-II)' was applied and two objective functions of 'Reliability' and 'Cost' of system were defined. Six design variables were set to related variables for system configuration. To verify the result of the optimization, the results of existing design and optimum design were compared in aspects of reliability and cost. With the optimization method proposed from this study, it was possible to derive the reliable and economical design of the ESD system.

1 Introduction

1.1 Motivation

As more offshore plants are installed around the world, more accidents related to the offshore plant areoccurring. Since 1995, the number of accidents related to the offshore plants for oil production has reached several hundred a year and a lot of people have been also injured or lost their lives[1]. Especially, most of offshore plants which are designed to drilling, production, retrieve, refine the oil are closely related to the flammable hydrocarbon gas in high temperature and high pressure.

Since the accident in Piper Alpha[2], the offshore plant industries recognized importance of safety from accident of hydrocarbon and fire/explosion in offshore plant. So, to reduce the many accidents and risks, various attempts have been made such as rule revision and creation of safety division. UK put the onus on the operator to identify the major hazards and to reduce risks with The Offshore Safety Case regulations[3]. The HSE (Health and Safety Executive) also created the 'Offshore Safety Division' and discussed the revision or verification of rules for safety. The NPD (Norwegian Petroleum Directorate) founded 'Regulations relating to management in the petroleum activities' in 2001 for safety [4].

[a] Corresponding author : scshin@pusan.ac.kr

There are a lot of approaches to satisfy safety in offshore plant. One is to reduce the 'Probability' of accident from human and organizational factors, system failure, natural disaster, etc. The other is to reduce 'Consequence' severity of such an event when it occurs with visual alarms, fire suppression system or a process shutdown [5]. From these aspects, the ESD system is very important to reduce 'Consequence' of accident as shutdown release of hazardous material. If the ESD system doesn't work and fail to shutdown when there is release of hydrocarbon in offshore plant, this failure could cause of fire/explosion disaster. So the ESD system is required to design with high reliability to avoid failure in dangerous situation.

From reliability aspects, there are two international safety authorities governing SIL (Safety Integrity Level), IEC (International Electrotechnical Commission) 61508 and IEC 61511. 61508 governs the functional safety of electrical, electronic and programmable electronic safety systems e.g. Production Inflow Control Devices (ICDs). It is applied across all industries and IEC 61511 governs the functional safety of safety instrumented systems and itis applied in the process industries. In 2000 year in Norway, OLF (The Norwegian oil industry association) tried to issue a guideline on the application of IEC 61508 and IEC 61511 in the Norwegian Petroleum Industry [6]. OLF also has defined the procedure and requirements of the ESD for offshore plant in their "Technical Safety" of 'NORSOK STANDARD S-001' [7]. DNVestablish 'OFFSHORE STANDARD DNV-OS-E201: Oil and Gas Processing Systems' and to provide an internationally acceptable standard of safety for hydrocarbon production plants and LNG processing plant by defining minimum requirements for the design, materials, construction and commissioning of plant[8].

1.2 A literature review

For high reliability of the offshore system, reliability analysis is necessary in the early stage of design. There are a lot of domestic and overseas studies related to the reliability analysis. As for the overseas studies, there was a research that suggested the simplified technique of reliability analysis and applied it to the offshore plant mooring system for the optimal [9]. There was also a study on the fatigue reliability analysis in the structure based on the analysis of various scenarios related to the structural fatigue for the extension of lifetime of the offshore plant [10]. But they are focused on structural or fatigue reliability of system. It is differ from functional safety of electrical, electronic, programmable electronic safety-related systems or safety instrumented systems for the process industry sector such as the ESD system.

As the overseas study directly related to the reliability analysis of the ESD system, FTA (Fault Tree Analysis) was used to define the failure rate of system component as the lower level and enhance the reliability of the system based on the HAZOP (HAZard and OPerability)[11]. SINTEF (Norwegian: Stiftelsen for industriell og teknisk forskning)studied reliability of subsea BOP systems for deepwater application[12]. Detailed failure statistics for the various BOP systems were analyzed and presented in the US GOM OCS (Outer Continental Shelf).Ram K. et al studied impact of reliability or the number of emergency shutdown devices on flare relief system and analyzed related factors for sizing of individual relief valves protecting equipment or process or system [13].This paper highlighted several concerns such as standards, reliability, safety and offers practical advice to those facing relief system design decisions.A.C. Torres-Echeverrı́a et al studied about multi-objective optimization for safety instrumented systems of chemical reactor system with three objective functions reliability, STR (Spurious Trip Rate) and cost[14].Theyappliedthe reliability modelstooptimizationofdesignandtestingof safety instrumented systems. The models for optimization have been integrated, together with a Life cycle Cost model, as objective functions in to a multi-objective genetic algorithm.FaresInnal et al also studied safety and operational integrity evaluation and design optimization of safety instrumented monitoring systems with two objective functions reliability and STR [15].

In domestic studies, there was a study about design of the flight control system. Reliability of the system was analyzed and the method of improving reliability through simulation was proposed [16].There was also another research in the field of fire prevention. The design of the system can be verified whether it is proper to the SIL through the reliability analysis of fire/explosion safety device of Ethyl Benzene process [17]. In offshore industry, Bae J. H. et al performed reliability analysis of the ESD for supporting design of LNG bunkering [18].

This study was focused on not only method of design optimization for offshore process but also practical design by selecting ESD products on the market.Totally 22 types of ESD components were investigated from valve companies and online. In order to design closer to practical system, Existing system 'Heidrun (TLP)', has been operating in Norwegian Sea since 1995, was selected to optimize design of ESD system and to compare its results.The multi-objective design optimization was performed with two objective functions of 'Reliability' and 'Cost'. 'Reliability'is based on PFD (Probability of Failure rate on Demand)values from reliability analysis and 'Cost'is composed of purchase cost, proof test cost, loss of production and etc. Design variables were set to six practical variables for configuration of system.To verify improvement of the design, the results of Heidrun design and optimum design was compared in aspects of reliability and cost.

2 The Emergency ShutDown system

In this thesis, the ESD system of 1st separation system in TLP at Heidrun oil field was selected for target system because it could be applicable more practically for optimization.The 1stseparation and related line has high pressure and temperature conditions with hydrocarbon material. It could havehigh risks of fire/explosion accident. So, these separation systems are required to controlled and monitored in all process functions on the topsides as well as Fire & Gas and the ESD for the entire FPSO. The P&ID (Piping & Instrument Diagram)of the ESD system is as shown in Figure 1[19].

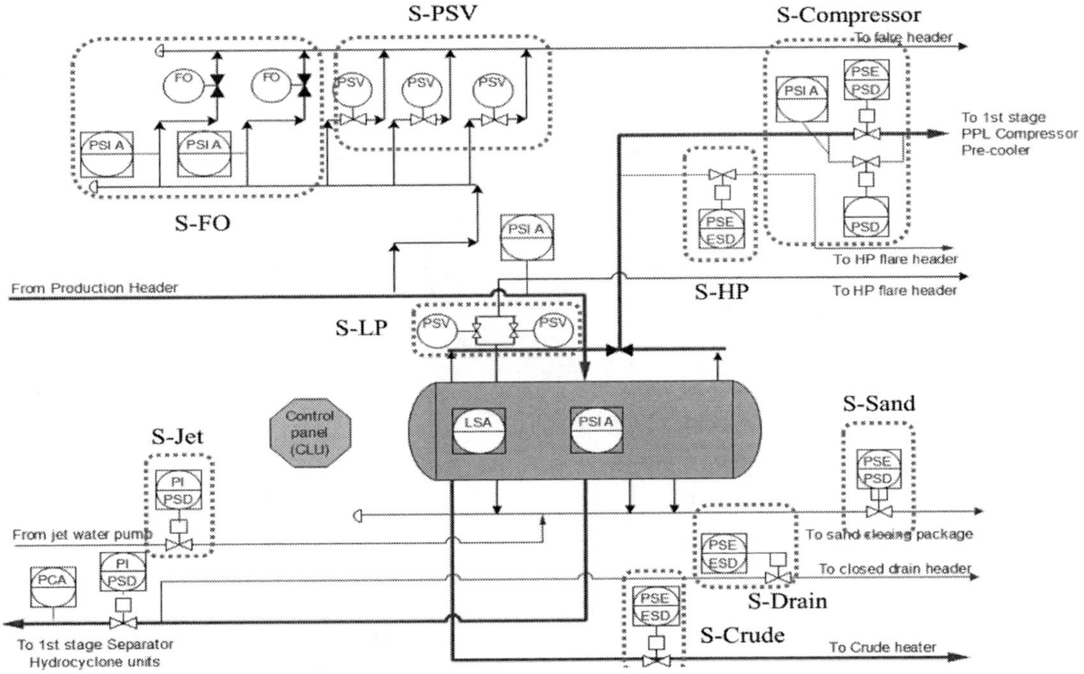

Figure 1.1 stseparation system with the ESD system [19].

Rectangular with dot line in Figure 1 presents the component of ESD system such as PSV (Pressure Safety Valve), ESD valve, PSD (Pressure ShutDown) valve, FO (Flow Orifice), PSI A (Pressure Safety Indicator/Alarm) and PSE (Pressure safety sensor). Equipment is expressed in P&ID with symbol and identification letters defined from American National Standard 'Instrumentation Symbols and Identification' [20]. Control panel (CLU: Control Logic Unit) is connected all of the ESD components.

3 Reliability Analysis

Reliabilityis defined by IEC 50 (191) as'the ability of an entity to perform a required function under gi ven conditions for a given time interval'and it is usually expressed in failure rate, MTTF (Mean Time To Failure), SIL (Safety Integrated Level) and etc. [21].

To perform reliability analysis for the ESD system, shutdown procedure is as follows;

1. If overpressure is detected by the sensors during separating operation, the main pump related to the 1stseparator is stopped immediately.
2. The PSD/ESD control logic send shutdown signal to final elements.
3. Final elements shutdown system to prevent further accidents from occurring.

3.1 PFD and Failure scenarios

Nine failure scenarios of overpressure were defined for reliability analysis (PFD calculation) as referred to 'Component structure'[22] and The Norwegian Oil Industry Association[6].The PFD of the E/E/PE safety-related system is determined by calculating and combining the average probability of failure on demand for all the subsystems which provide protection against a hazardous event [22].

The failure scenario 'Flare FO' is related to the failure of two flow orifices, two pressure safety indicators installed in the line to flare header and CLU for control. If there is the overpressure in line, CLU should order to open the flow orifice. Once one of two flow orifices operates normally in failure situation, this scenario is success as shown in Figure 2. In similar way to define scenarios such as 'Flare FO', theother eight scenarios were defined as shown Figure 3to Figure 10.

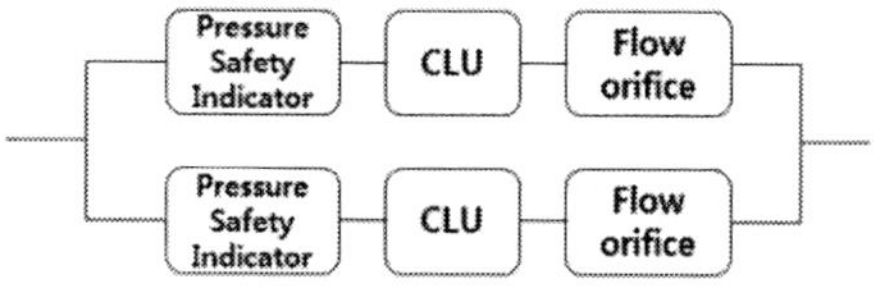

Figure 2. Scenario - Flare FO.

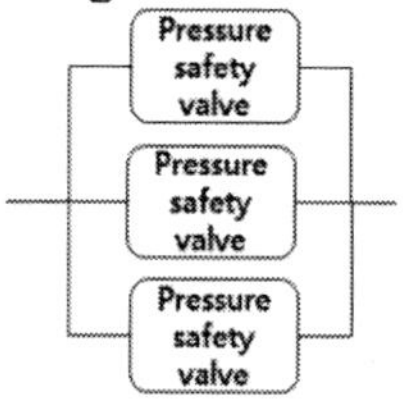

Figure 3.Scenario - Flare PSV.

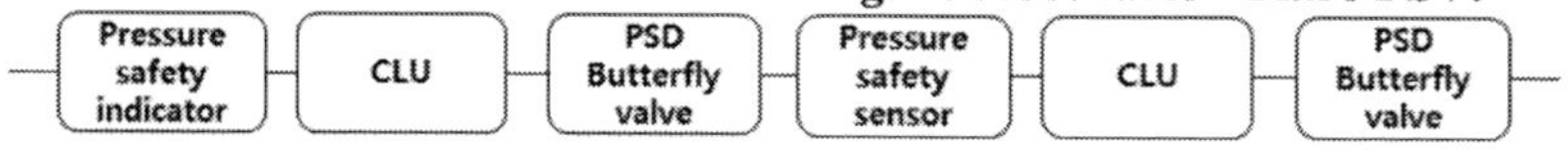

Figure 4. Scenario – Compressor

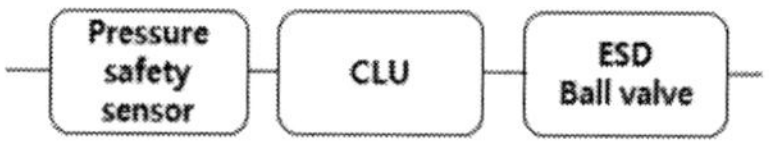

Figure 5. Scenario – HP (High Pressure) flare header

Figure 6.Scenario - Sand cleaning

Figure 7. Scenario - Drain header

Figure 8.Scenario - Crude heater

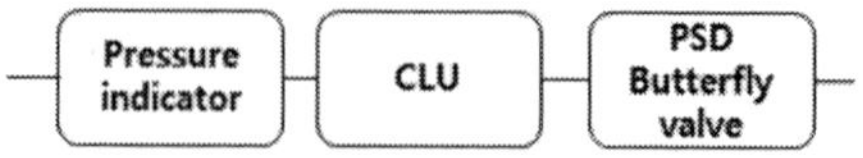

Figure 9. Scenario - Jet water pump

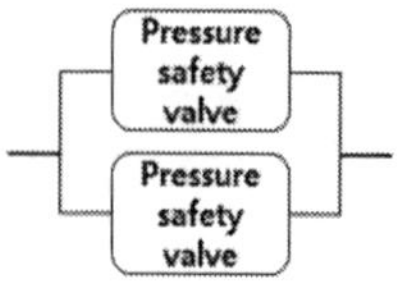

Figure 10. Scenario – LP (Low Pressure) compressor

3.2 Calculations of PFD and SIL

For, reliability analysis, failure data and MTTR (Mean Time To Repair)were referred from 'OREDA (Offshore and Onshore Reliability Data) 2009'[23]. From nine failure scenarios with failure data of components, PFD and SIL were calculated as shown in Table 1.

Table 1. The results of reliability analysis (PFD and SIL).

Scenario	Component	Type	Failure rate (per10E+6hours)	MTTR (hours)	PFD	SIL	RequiredSIL
S-FO	Sensor	PSI	4.20E-07	4	1.21E-03	SIL 2	SIL 2
	Logic unit	CLU	2.85E-05	6			
	Final element	FO	4.23E-06	8			
S-Flare PSV	Final element	PSV	8.47E-06	8	1.48E-04	SIL 3	SIL 2
S-Compressor	Sensor	PSI	4.20E-07	4	1.73E-02	SIL 1	SIL 2
	Logic unit	CLU	2.85E-05	6			
	Final element	PSD	1.90E-05	17			
	Sensor	PSE	4.10E-07	4			
	Logic unit	CLU	2.85E-05	6			
	Final element	PSD	1.90E-05	17			
S-HP flare header	Sensor	PSE	4.10E-07	4	9.89E-03	SIL 2	SIL 2
	Logic unit	CLU	2.85E-05	6			
	Final element	ESD	2.58E-05	16			
S-Sand cleaning	Sensor	PSE	4.10E-07	4	9.89E-03	SIL 2	SIL 2
	Logic unit	CLU	2.85E-05	6			
	Final element	ESD	2.58E-05	16			
S-Drain	Sensor	PSE	4.10E-07	4	8.66E-03	SIL 2	SIL 2
	Logic unit	CLU	2.85E-05	6			
	Final element	PSD	1.90E-05	17			
S-Crude heater	Sensor	PSE	4.10E-07	4	9.89E-03	SIL 2	SIL 2
	Logic unit	CLU	2.85E-05	6			
	Final element	ESD	2.58E-05	16			
S-Jet water pump	Sensor	PSI	4.20E-07	4	8.66E-03	SIL 2	SIL 2
	Logic unit	CLU	2.85E-05	6			
	Final element	PSD	1.90E-05	17			
S-LP compressor	Final element	PSV	8.47E-06	8	1.48E-04	SIL 3	SIL 2

From, the results of reliability analysis, scenario 'S-Compressor' has lowest SIL 1 and scenarios 'S-PSV', 'S-LP' have high SIL 3. Except these 3 scenarios, allscenarios have SIL 2. Even if 'S-FO' has already SIL 2, it has more chance to reduce cost with higher PFD value in SIL 2 range.If the ESD system for 1stseparation system in offshore plant is required to minimum SIL 2 as referred from The Norwegian Oil Industry Association[6], 'S-Compressor' is needed to improve design to meet SIL 2 from SIL 1, while S-PSV' and 'S-LP' are needed to simplify design to make SIL 2 from SIL 3 for reducing cost.

4 The design optimization of the ESD system

4.1 Definition of Optimization problem

The purpose of this design optimization is to find design variables that make minimum value of objective function. It means optimized design has high reliability with reasonable cost for the ESD system. NSGA-II is selected for optimization algorithm.

4.1.1Objective function

- Objective function 'Reliability'

Objective function of 'Reliability' (f_1) is estimated from each scenarios' as equation (1).

$$minimize\ f_1 = PFD \tag{1}$$

- Objective function 'Cost'

Objective function of 'Cost'(f_2) is calculated from 'Product cost', 'Replace cost', 'Proof test cost' and 'Loss of production' of the ESD systems on the following equation (2).

$$minimize\ f_2 = C_{Product} + C_{Replace} + C_{Proof\ test} + Loss_{Production} \tag{2}$$

- $C_{Product}$ is product price of sensors, logic unit and final element for installation at first time.
- $C_{Replace}$ is cost of replacement during lifetime thatdepends on MTTF.
- $C_{Proof\ test}$ is calculated based on times of proof test during lifetime, test cost of labor[24] for one equipment and number of equipment.
- $Loss_{Production}$ is loss of production from downtime duringproof test[25]. It estimated with WTI crude oil price $57 (April 17, 2015), production '65,000bbl/day' at 'Heidrun' oilfield[26] and test time '1 hour'[27].

4.1.2 Design variables

From a reliability point of view, system is generally consist three parts; sensor, logic unit and final element. As shown in Table 2, six design variables were set to the number of redundancy at each part, type of sensor and final element, proof test interval.Database for design space was created including information of products as MTTF and price. Eight types of sensors and fourteen types of final elements were investigated from brochure of product[28] andonline market[29]. Failure data is referred to 'OREDA 2009' data for sensors, logic unit and final elements.

Table 2. Design variables and space.

Design variable		Unit	Range
The number of redundancy - sensor		Number	0, 1, 2
The number of redundancy - logic unit		Number	0, 1, 2
The number of redundancy - final element		Number	0, 1, 2
Type of sensor		Type	1~8 (8 types of products)
Type of final element	For blowdown	Type	1~2 (2types of products)
	For shutdown	Type	1~12 (12 types of products)
Proof test interval		Year	1~3

4.1.3 Constraints

The topside process in offshore plant is not extremely dangerous such as nuclear plant or has not very severecondition such as deepwater subsea well operation. Therefore generally SIL 2 is proper for offshore topside process. The Norwegian Oil Industry Association[6] also suggested minimum SIL 2 for the ESD system related to separation system. Constraints were set to SIL 2 and it has range of $10^{-3} \leq \text{PFD} < 10^{-2}$ by PFD value.

4.2 The results of the optimization

4.2.1 'S-FO' - Blowdown operation

Population of NSGA-II was set to 40, generation was 1,000 andcalculation time was 9.4s for optimization. Figure 11 is Pareto-frontier results from the optimization of scenario 'S-FO'. In this study, we focused on optimum design which has minimum cost in SIL 2. This means among the alternatives which satisfied SIL $2(10^{-3} \leq \mathrm{PFD} < 10^{-2})$, lowest cost alternative 'FO'could be chosen as shown in Figure 11.For 'S-FO' - Blowdown operation in 'To flare header' line, It should have equipment for blowdown system such as flow orifice. So, type of the final element in 'S-FO' was fixed to flow orifice and optimization was performed with the other design variables type of sensor, the numbers of redundancies and proof test interval.

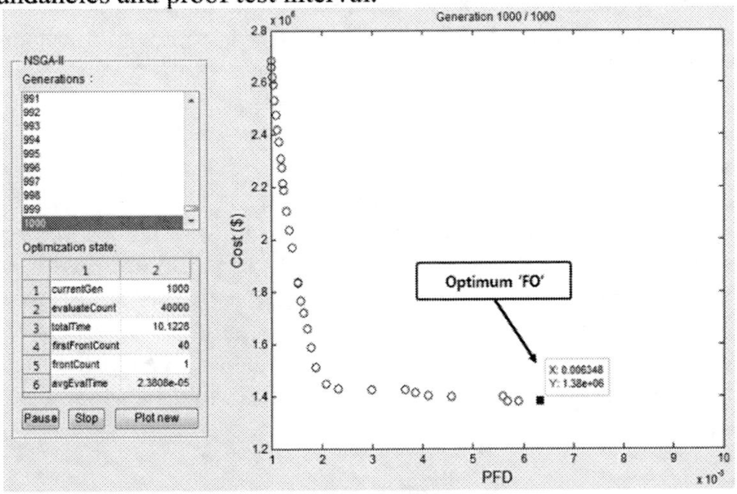

Figure 11. Pareto-frontier of scenario 'S-FO' and optimum alternative

Details of alternative 'FO' (0.006348, 1.38e+6) are as shown in Table 3.

Table 3. Optimum alternative of 'S-FO'

Design variable	**Value**	**Details**
Sensor	0	No redundancy
Logic unit	1	1 redundancy
Final element	0	No redundancy
Type of sensor	2	Pressure indicator 'P Series'
Type of final element (blowdown)	2	Flow orifice
Proof test interval	3.000	3 years

4.2.2Summary of the results include other eight scenarios

The total results and comparisons of optimization results to Heidrun system are as shown in Table 4.

Table 4. The total results and Comparison of optimization results.

Heidrun				**Optimum**		
Scenario	PFD	SIL	Cost($)	PFD	SIL	Cost ($)
S-FO	0.0012	SIL 2	4,220,513	0.0063480	SIL 2	1,379,840
S-Flare PSV	0.0001	SIL 3	4,035,682	0.0048284	SIL 2	1,384,680
S-Compressor	0.0173	SIL 1	4,220,463	0.0098441	SIL 2	1,452,460
S-HP flare header	0.0099	SIL 2	4,047,008	0.0065769	SIL 2	1,383,100
S-Sand cleaning	0.0099	SIL 2	4,047,008	0.0065769	SIL 2	1,383,100
S-Drain	0.0087	SIL 2	4,047,008	0.0065769	SIL 2	1,383,100
S-Crude heater	0.0099	SIL 2	4,047,008	0.0065769	SIL 2	1,383,100
S-Jet water pump	0.0087	SIL 2	4,047,008	0.0065769	SIL 2	1,383,100
S-LP compressor	0.0001	SIL 3	3,975,550	0.0044857	SIL 2	1,353,610
Total cost			36,687,248			12,486,090

Every scenario is optimized to meet the minimum SIL 2 and total cost of final design also decreased $24,191,186 from origin design.

4.3Discussion

As shown in Figure 12, all PFD values of scenarios are in the range of SIL 2 and this means they satisfied the required reliability through the optimization.Although SIL of the scenario 'S-FO' is the same as SIL 2 before the optimization, PFD is increased up to about 0.005 for reducing cost by design modification. In case of the scenario 'S-FO', redundancy was removed and another element among the database that has lower PFD was selected to reduce the cost of system in SIL 2.PFD of'S-PSV' and 'S-LP' scenarios were also increased and their SIL was degraded to SIL 2 from SIL 3 to reduce the cost. To design system with higher reliabilityneeds more cost because they need generally high quality products and complex system.But from the results PFD and cost as shown Figure 12, it was possible to improve reliability andreduce cost simultaneously.

PFD values of eight scenarios except 'S-Compressor' could not reached close to boundary of SIL 2 and SIL 1 as shown in Figure 12. It means they could have still more possibilities of improvement with reduction of cost. Despite convergence of optimization in this study, to reach near the ideal optimum point 'boundary of SIL 2 and SIL 3' was difficultbecause there were discrete design variables such as type of element and the number of redundancy. One of the methods of improve the result of optimization is to adding various elements for increasing database in order to make design space almost continuous.

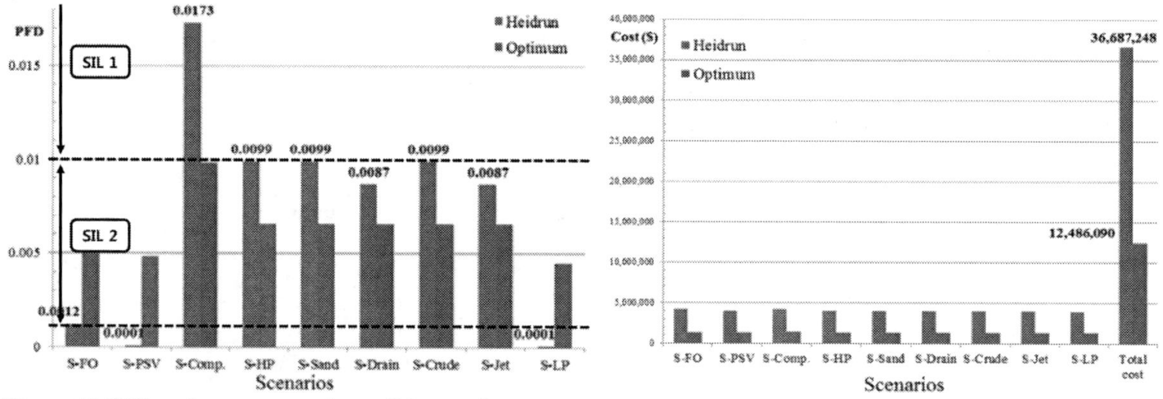

Figure12.PFD and cost comparison of the results.

As comparison of design variables of Heidrun and optimum as shown in Table 5, all of test intervalsare increased and all structures of S-L-F (Sensor-Logic unit-Final element) are changed. Number in S-L-F column of Table 5 means the number of element in each scenario. The number of sensor and final element are modified to the same as one except 'S-PSV' and 'S-LP'. It seems they tried to decrease the number of redundancy for reducing cost of each scenario.From the results of 'Type (sensor)' in Table 5,'1: Pressure safety indicator' and '2: Pressure safety Sensor'are considered suitable for the ESD system in this study.

Table 5. Comparison of design variables of Heidrun and optimum.

Operation mode	Scenario	Heidrun						Optimum					
		S-L-F (structure)			Type (sensor)	Type (final element)	Proof test interval	S-L-F (structure)			Type (sensor)	Type (final element)	Proof test interval
Blowdown	S-FO	2	2	2	1	2	1.000	1	2	1	2	2	3.000
	S-PSV	0	0	2	-	1	1.000	0	0	1	-	1	3.000
Shutdown	S-Comp.	1	1	1	1	3	1.000	1	2	1	2	3	2.999

		1	1	1				1	2	1			
	S-HP	1	1	1	2	8	1.000	1	2	1	1	2	3.000
	S-Sand	1	1	1	2	12	1.000	1	2	1	2	2	3.000
	S-Drain	1	1	1	2	8	1.000	1	2	1	2	2	3.000
	S-Crude	1	1	1	2	8	1.000	1	2	1	2	2	3.000
	S-Jet	1	1	1	2	12	1.000	1	2	1	2	2	3.000
Blowdown	S-LP	0	0	2	-	1	1.000	0	0	1	-	1	3.000

From Table 5, the number of logic unitforsix scenarios 'S-Comp.', S-HP', 'S-Sand', 'S-Drain', 'S-Crude', 'S-Jet' are increased to two from one. We can estimate that this reason from graph of PFD comparison as shown in Figure 12. All of six scenarios' PFD values are decreased and this means redundancy of logic unit was be used for reduction of PFD.

5 Conclusions

In this study, following were carried out in order to attain final goals.

- Reliability analysis of the existing ESD system foroffshore process was performed with defined scenarios and failure data.
- The multi-objective design optimization was performed with defined two objective functions of 'Reliability' and 'Cost'. Six design variables and 'SIL 2' constraints were defined. Optimum design was selected from Pareto-frontier and it satisfied both reliability SIL 2 and cost reduction.
- In order to designcloser to practical system, existing system was selected to optimize design of ESD system. Database for design space was also created including information of product on the market.

With these results, more practical method of design optimization was proposed for the ESD system of offshore process and it could be applied to other similar process.One of the methods of improve the result of optimization is to adding various ESD elements for increasing database and makes design space almost to be a continuous.

Acknowledgements

This work was supported by the National Research Foundation of Korea (NRF) grant funded by the Korean government (MEST) through GCRC-SOP (No. 2011-0030671).

References

1. HSE, OFFSHORE INJURY, ILL HEALTH AND INCIDENT STATISTICS 2010/2011 (2011)
2. M. Elisabeth Pate-Cornell, Risk Analysis, **13**, 2 (1993)
3. HSE, A Guide to the Offshore Installations (Safety Case) Regulations (1992).
4. NPD,Regulations relating to management in the petroleum activities(2001)
5. CahillJ., Differences in Fire & Gas Systems and Emergency Shutdown Systems (2011)
6. OLF, *APPLICATION OF IEC 61508 AND IEC 61511 IN THE NORWEGIAN PETROLEUM INDUSTRY*(OLF 070, 2004)
7. OLF, *NORSOK STANDARD S-001 Technical safety* (OLF, 2008)
8. DNV, *OFFSHORE STANDARD DNV-OS-E201: Oil and Gas Processing Systems*(DNV, 2013)
9. Emad MousaviM., GardoniP., Marine Structures, **36** (2014)
10. GholizadaA., GolafshaniA. A., AkramiV., Ocean Engineering, **46** (2012)
11. DragffyG., RESS, **61** (1998)

12. SINTEF, *Reliability of Subsea BOP Systems for Deepwater Application, Phase II DW Unrestricted version*(SINTEF, 1999)
13. Ram K. Goyal, Essa G. Al-Ansari, JLP, **22**, 35 (2009)
14. Torres-EcheverriaA.C., MartorellS., ThompsonH. A.,RESS, **106** (2012)
15. Innal F., Dutuit Y., Chebila, M., RESS, **134** (2015)
16. Kim, S.S., Ph. d thesis of INHA Univ., (2011)
17. Ko, J.S., Kim, H., Lee, S. K., Fire Science and Engineering, **20**, 3 (2006)
18. Bae, J. H., Shin, S. C., Kim, S. Y., Proceedings of Naval Architects of Korea, pp.965-970 (2014)
19. ConocoPhillips, *Safety Shutdown Systems, additional technical requirements* (ConocoPhillips, 2007)
20. ANSI/ISA, *Instrumentation Symbols and Identification*(ISA, 1992)
21. IEC, *International Vocabulary, Chapter 191: Dependability and Quality of Service*(IEC 50 (191), 1991)
22. IEC, *IEC 61508-6, Part 6: Guidelines on the application of parts 2 and 3* (IEC, 1997)
23. SINTEF, *OREDA Offshore Reliability Data 5th Edition*(SINTEF, 2009)
24. GrossR. E., HarrisS. P., RAMS 2008, pp.312-316 (2008)
25. DavidDr., SmithJ., *RELIABILITY MAINTAINABILITY AND RISK Practical methods for engineers 8th edition*(Butterworth-Heinemann, 2011)
26. Wikipedia, *Heidrun oil field*(Wikipeida, 2015)
27. Rausand M., Reliability of Safety-Critical Systems: Theory and Applications (WILEY, 2014)
28. Metso Automation, *ESD VALVE SELECTION GUIDE GENERAL ESD VALVE DEFINITION*(Metso Automation, 2005)
29. OMEGA Engineering inc., http://kr.omega.com/products.html (2015)

MATEC Web of Conferences **52**, 02004 (2016)
DOI: 10.1051/matecconf/20165202004

Product System Design – to Household Massage Design as An Example

Wang Huabin[1], Lin Qichao[2,a]

[1]*Industrial Design, South China University of Technology School of Design, University town, Panyu district, Guangzhou ,Guangdong Province, China*
[2]*Industrial Design, South China University of Technology School of Design, University town, Panyu district, Guangzhou ,Guangdong Province, China*

Abstract. Explain what is the system design and the applications of System design in the product design process. Using The whole idea and systems design methods to analyze the design of household hand massage, Use a chart image to explain that Household hand massage products in the influence of user and the environment , the influence of large system environment for the product. The use of components of the system to anatomy product design . Each system components has a link between and mutual correlation.

1 System Design and the Basic Framework of Product System Design

Product system design has two meanings, One is the design object is no longer isolated, but put it in the relevant system in consideration; Secondly, a single product also be regarded as a system made up by more than one unit, and in order to realize the product in the use of functional interchange ability, flexibility and infinite complementary.[1]

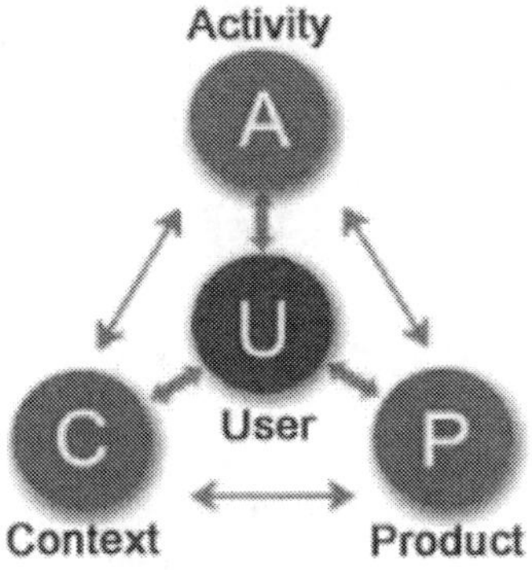

Figure 1-1. the basic framework of UACP product system design.

"Interaction design", points out that in the field of Product design related to industrial design, can use the User, Activity, Context and Product four elements to constitute the basic framework of UACP Product system design.[2] The product is designed by the Interactive system environment of PACT

[a] Corresponding author : littlepig199112@hotmail.com

posed organic, PACT is short for interactive system elements, People (People), behaviour (Activity), the product when using the scene (Context) and the integration of Technology in products (Technology).[3] The analysis of the nature of the product and function through elements of the system , emphasizes the system constitute harmony of elements, Emphasize the interactive function of product in the use of the environment, focus on the user and the product system experience in the process of information exchange. Products can be derived from two kinds of design patterns, "people-things" system," people-things - environment" system.

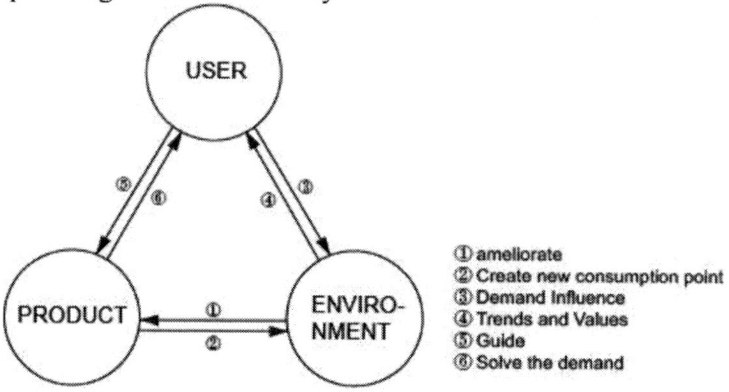

Figure 1-2. " people - things - environment " system model

2 The household hand massage products system

Household hand massage product is implicated in multiple systems analysis and design. Designers should think deeply products from the perspective of system design, design and develop Systematic solution with products as the core.

2.1 User

Use of the method of qualitative characters from the user research to determine the target user. Based on qualitative research, interviews, field observation and usability testing, creating characters, according to the user's goals, find some views and behavior model, subdivide each group to create a character. Unified the target group is the concept of fuzzy, tend to sit for a long time, believe and understand massage physical function, often feel body muscle tension and mental pressure of the crowd. Designing products by the characteristics of the user. products,behavior and situations can affect ideas and new knowledge of users.[4]

2.2 Behavior

User research methods observed user behavior, this type of users who sit for a long time don't like the use of complex means and methods, so the key to attract the users of this type are how to relieve pressure and muscle tension by massage and human-computer interaction.

2.3 Situation

After determine the target user groups, with the overall system design method to consider the use of the user. Products are communication between people and things in a situation and moment , for realizing people's needs, Products are to rely on the scene.

2.4 Product

The product is the core of the system components, including tangible products and intangible product (service). For the product of relieve stress, and sometimes the user need may not be the product itself, but the back of a product service. Product form of entity is broken, makes the product is not only physical properties can also as a bridge connecting people to achieve real communication.

In the application of home massage products in two kinds of system model

"people-things" system: household massage product properties by material function to the user, to meet their material needs.

" People - things - environment " system: People in contact with the product that have an effect on society and the surrounding environment , at the same time, people affected products, the environment also affect a person, actually is a cyclic process. Products in addition to meet the user's physiological needs need to meet the needs of psychology, psychological demand is closely linked with people in the environment. " people - things - environment " is a large system. The mutual influence between them. (as shown in figure 2-1)

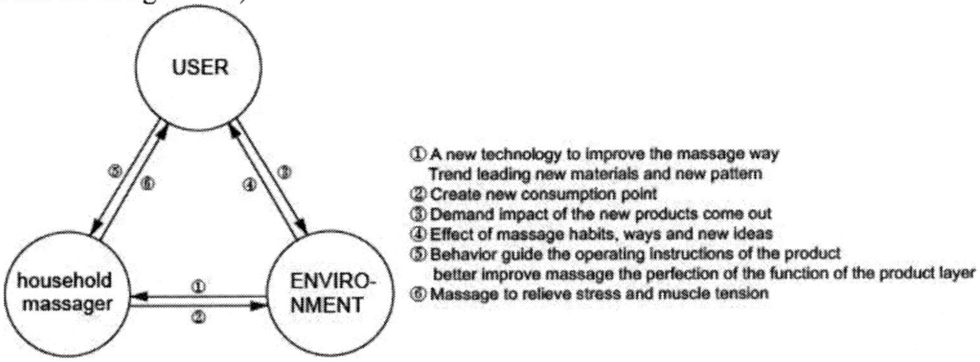

Figure 2-1. "household massage - User - the external environment " system model. The relationship between the system and influence each other

"Household massage - User - the external environment " system model.System link belong to each other between the universality of specificity, in view of the system with different products is the practical.

The entire product internal system and external system linked to each other, the relationship between the internal system is the product of the whole system, there are four basic elements. External system is the external environment, which include the existing technical environment, use of environment and economic environment, etc. Between them is through the whole system connect with each other, influence each other (as shown in figure 2-2)[5]

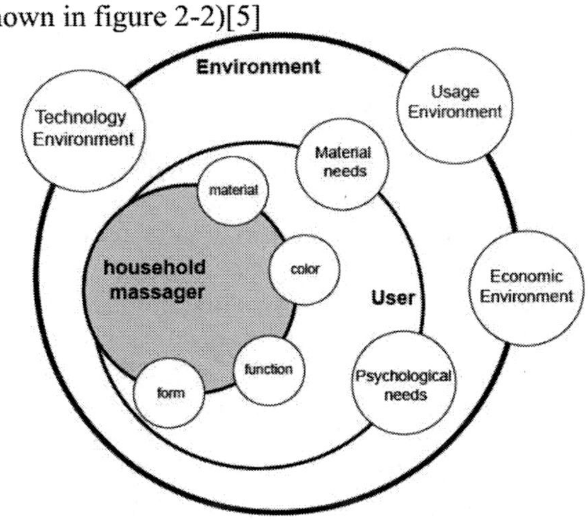

Figure 2-2. the relationship between the internal system and external system

3 Household Massage Products System Components, And Feedback In Research

Household system elements are: material elements, color elements, function, shape, etc. The early stage of the analysis of product by using the principle of the system design, research of the product elements and sums up their advantages is to master the whole industry situation and understand the design direction has important guiding role.[6]

Materials can be found through the actual investigation form elements, color elements, function, shape elements influence each other, interaction are shown in table 3-1

Table 3-1. household massage products system elements of network survey table

Form	material	color	function
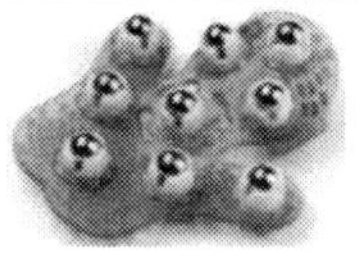	Food grade PVC POM Steel ball	Sky blue pink violet	1.360 degrees rotation platinum film steel ball, cooperate the design accord with human body type palm one-piece silicone floppy disks, giving a relaxing massage. 2. Massage bead can close to the muscles for all-round reciprocating rotary movement.
	External: frosted PVC	Blue with white Pink with white	1. The high frequency vibration massage (80 Hz, 4800 r/min), the contact stress points to highlight can dredge meridians, to reconcile blood, relieve stress. 2. Rotation can adjust at will massage strength, easy to operate.
	Food grade PVC	Warm gray	1. The systemic pressure point massage, high speed vibration, relieve muscle tension and ache, and promote blood circulation. 2. Provide additional side handle grip button is simple, easy to operate.

Material elements: as the product is rich, material limits for the product are gradually being broken. Now home massage products material is also in the implementation of new reform, mixing of different material, the use of different materials, the combination of soft and hard materials to the user the use effect of different, the use of the user behavior is more smooth and perfect.

Color: Color is an important language and elements in the product design. After a lot of experts and scholars of research confirms that color can affect the person's psychology. the color of household hand massage should be required to calm the user's psychological pressure, and also has the very good affinity.

Functional elements: function is the primary elements of products. "Not completely without any function of the object in the world, since its objective existence is inevitably has its reason to exist", the stand or fall of the function of a product design is often the most direct performance, meet the demand of the user functions in terms of product availability and ease of use is one of the main evaluation criteria. Household hand massage should be based on function of massage, different massage way also concerns of the functional factors, at the same time satisfy the basic function according to the previous research, to access to the user's access to survey method, seek the function of the user's real needs, such as heating, mobile massage and so on.

Form elements: form is the most direct the actual image of a product to users, is one of the most powerful factor to improve product added value. Household hand massage morphology according to the function to decide, according to the user's psychological needs. Previous research in the form of different products can realize different function, produce the advantages and disadvantages of product evaluation, can very good lead the implementation of the new, improved product form.[7]

4 Summary

Product design is to have the idea of the overall consideration of each system, each element in the system and the relationship between each system. The whole is greater than the part, the overall could not be isolated and focus on a part of the system .we should integrally look at a part need to meet the consumer demand of whole product design. Product system, the user's system, external environment system sometimes appears to be no relationship between them, actually have a deep affinity.

References

1. Jiang Wen. System design---A Way for Combining the Function of Products. Chinese Journal of Engineering Design. 28(2001)
2. Zhou Fei. Deng Rong. Li Shiguo. Research on Fuzzy on the Product Interaction Design PACKAGING ENGINEERING 34 39-40（2013）
3. David Benyon,. Phil Turner. Susan Turner. Designing Interactive Systems. Pearson Education Limited. 65-72 (2005)
4. Don Norman. Emotional Design .Basic Books (2005)
5. Rao Heng. Research On System-product Design Of Tea set Based On The SpD Theory [D].Guangdong University of Technology. 34-37 (2012)
6. Cai Ke. The Element Analysis of Product System Design [D]. Wuhan University of Technology 10-18(2008)
7. Rao Heng. Research On System-product Design Of Tea set Based On The SpD Theory [D].Guangdong University of Technology. 34-37 (2012)

MATEC Web of Conferences **52**, 02005 (2016)
DOI: 10.1051/matecconf/20165202005

Effect of Local Constraint on Driving Torque of Driving Mechanism for Half-rotating Wing

Huixing Wang, Xiaoyi Wang[a], Han Qiu, Fuqiang Chen, Zhizhen Qiu and Jianhua Shan

School of Mechanical Engineering, Anhui University of Technology, Ma'anshan, 243002, P.R.China

Abstract: Half-rotating wing (HRW) is a new kind of flapping wing system with rotating-type flapping instead of oscillating-type flapping. For the complex mechanical transmission and high weight which is not suitable for flight of traditional half-rotating mechanism (HRM).The simplified HRM with local constraint was proposed as the driving mechanism of HRW in this paper. The performing process and characteristics of local constraint was further given. According to contact state between moving components, the performing process of local constraint was divided into two stages. To analyze the influence of local constraint on the driving torque exerted in the crank of driving mechanism, the model of driving torque under different stages were respectively established. Based on the model, the change curve of driving torque was obtained through simulation by MATLAB, which showed that local constraint had little influence on driving torque. The results could provide guidance for the choice of motor of HRW and optimization of dynamics.

1 Introduction

Flapping-wing air vehicle had become research hotspot due to its excellent maneuverability and hovering ability beyond traditional fixed wing and micro rotary wing[1].Researchers superimposed two same sets of crank-rocker mechanisms to create the Micro Bat[2].Madangopal et al designed the Sparrow by combining slider-crank mechanism with double-slider mechanism[3].Wei Zhen et al invented the parallel crank-rocker flapping mechanism[4].However, these mechanisms would produce huge inertia force due to high frequency oscillation, which limited its operating frequency and the scope of work.

During the study of bionic machinery, it could be found that animals' movement was essentially asymmetric. A kind of single degree mechanism named HRM was proposed as the driving mechanism for HRW, which could achieve flapping just by asymmetric rotating. However, HRM driven by gear or chain[5-7] was not suitable for flight due to the complex mechanical transmission and high weight for which the simplified design was needed. To ensure that the movement of HRM is accurate in a given time, a temporary constraint, called local constraint, was added artificially in the process of simplification. On the other hand, torque on the driving mechanism would be changed when local constraint worked with and without. Therefore, mechanical model was established and change of driving torque was determined. Both of these were guidance for the choice of motor of HRW and optimization of dynamics.

[a]Corresponding author : wangxy@ahut.edu.cn

2 Driving mechanism of HRW and local constraint

2.1 Driving mechanism of HRW

The sketch of HRW was shown in Fig.1(a), whose driving mechanism was designed based on the principle of HRM. HRM was consist of crank and half-rotating rod, as shown in Fig.1(b).Half-rotating rod would turn half round anticlockwise when crank turned one round anticlockwise. However, it was hard to realize the moving rule of HRM according to this concept for the mechanism would interfere when rotation center C moved to stagnation point H. Therefore, one edge of the crank was designed as shaft with groove. Furthermore, positioning slider was set at the location of point H to make sure HRM had accurate moving rule. The combination of positioning slider and shaft with groove could realize simplified design of HRM in Fig.1(c). Herein, point C was the center of shaft, point H denoted the rotation center of positioning slider.

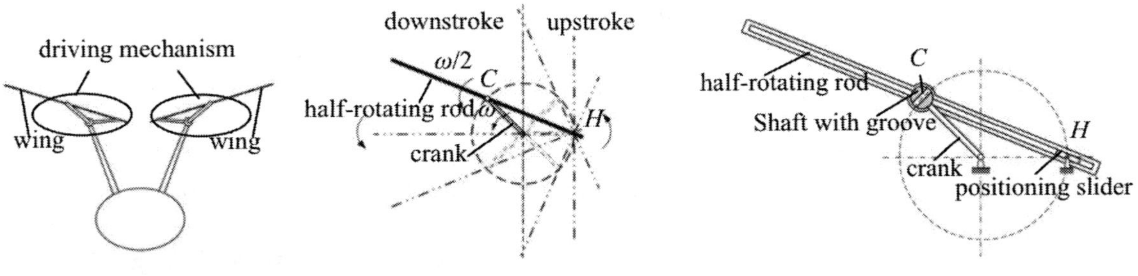

(a) Aircraft using HRM (b) Principle of HRM (c) The simplified HRM

Figure 1. HRW aircraft and its driving mechanism

2.2 Working principle of local constraint

In the movement of driving mechanism of HRW, when the center of positioning slider was not in shaft with groove, the half-rotating rod would rotate according to half-rotating law with constraint from the center of positioning slider. After center of positioning slider entered the groove of shaft, as shown in Fig.2(a), positioning slider could not rotate freely due to the constraint from arc-shaped groove at the interior of groove. Meanwhile, half-rotating rod would rotate with positioning slider due to the constraint at the exterior of groove. This constraint made up of positioning slider, shaft with groove and half-rotating rod, functioned transiently and locally, which was called local constraint. Fig.2 showed the relative position of positioning slider, shaft with groove and half-rotating rod with local constraint at different time.

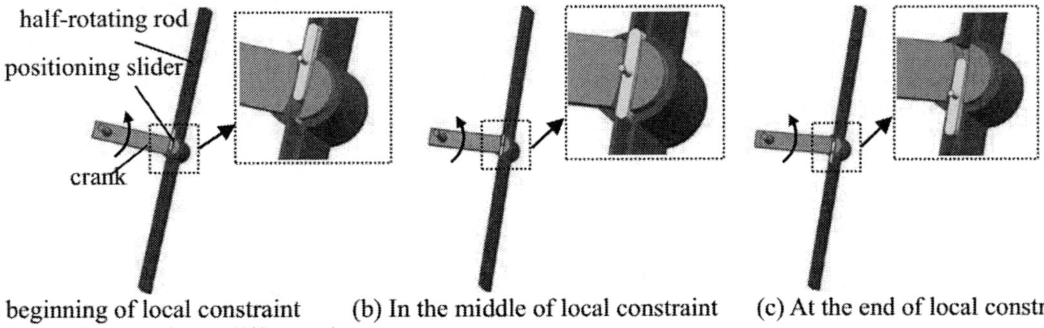

(a) The beginning of local constraint (b) In the middle of local constraint (c) At the end of local constraint

Figure 2. Local constraint at different time

3 Model of driving torque on driving mechanism for HRW without local constraint

Positioning slider would not affect the movement of half-rotating rod without local constraint. Fig.3 showed the simplified model of driving mechanism for HRW without local constraint. Herein, *OC* denoted crank, *AB* represented half-rotating rod. Blade of HRW and half rotation rod were fixed. The force from air on the blade was a distributed force system, which could be simplified as principal vector F_R and principal moment M_C at any instantaneous. The force on crank could be decomposed into X_O horizontal and Y_O vertical directions.

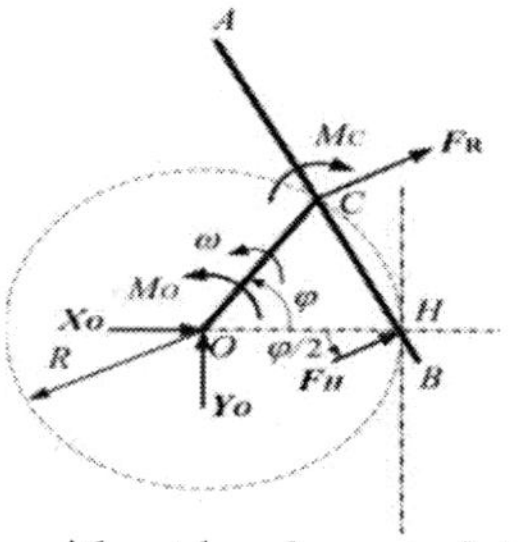

Figure 3. Force analysis of driving mechanism without local constraint

Model of driving torque M_o could be calculated by combining $\sum M_o = 0$ with $\sum M_C = 0$.

$$M_O = \frac{1}{32} C_D \rho h \omega^2 (16R^4 \sin^4 \frac{\varphi}{2} + 24R^2 a^2 \sin^2 \frac{\varphi}{2} + a^4) \quad (1)$$

Where C_D was drag coefficient of blade which could be obtained by experiment. ρ was air density. h was width of blade. a was length of blade. R was length of crank. ω was angular velocity of crank.

4 Model of driving torque on driving mechanism for HRW with local constraint

4.1 Different stages under local constraint

Inside contact stage was the first stage, which was started from the beginning of local constraint to the coincidence of point *C* and point *H*. The relevant components of driving mechanism under this stage was shown in Fig. 4(a). Herein, F_1 was the force on positioning slider excited by half-rotating rod, F_2 was the force on positioning slider excited by the inside arc of shaft with groove, F_H denoted the constraining force on positioning slider shaft excited by bracket. It could be seen clearly that, the inside surface of positioning slider and half-rotating rod were contacted with each other, and the inside surface of positioning slider contacted with the inside arc of shaft with groove.

Outside contact stage was the second stage, which started from the separation of point *C* and point *H* to the end of local constraint. The relevant components of driving mechanism under this stage was shown in Fig. 4(b). Herein, F_1 was the force on positioning slider excited by half rotating rod, F_2 was the force on positioning slider excited by the outside arc of shaft with groove, F_H was the constraining force on positioning slider shaft excited by bracket. it could be found from Fig. 4(b) that the outside surface of positioning slider and half-rotation rod were contacted with each other, and the outside surface of positioning slider contacted with the outside arc of shaft with groove.

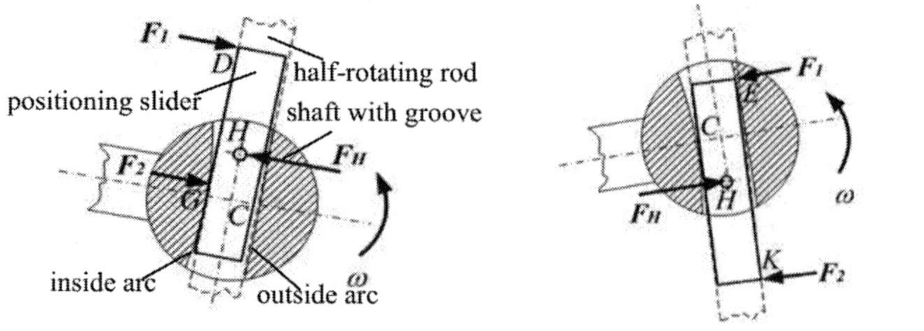

(a) Inside contact stage
(b) Outside contact stage

Figure 4. Different stages of local constraint

4.2 Driving torque model under different stages

This article was focused on the calculation of the driving torque under inside contact stage. As space is limited, other similar derivation was omitted, and the results were given directly.

4.2.1 Driving torque model under inside contact stage

The contact state of driving mechanism was shown in Fig.4(a).The force F_2, which passed through the center of inner arc, was closely perpendicular to the half-rotating rod. So the force system acting on the positioning slider could be regarded as a plane parallel force system.

(1) Firstly, half-rotating rod was regarded as research object and force analysis was shown in Fig.5

The groove of half-rotating rod and positioning slider were represented by solid line and dotted line respectively. $F_1^{'}$ was the force on half-rotating rod excited by the positioning slider and it could be calculated as follow.

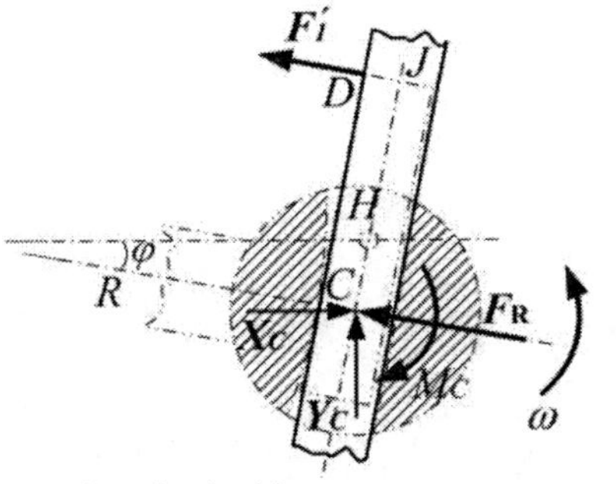

Figure 5. Force analysis of half-rotating rod under inside contact stage

$$F_1^{'} = \frac{M_C}{2R\sin\frac{\varphi}{2} + b} \tag{2}$$

Where b was half the length of the positioning slider and M_C could be obtained from literature [5].

(2) secondly, positioning slider was regarded as research object and force analysis was shown in Fig.6

Herein, line HE, which was tangent with circle O_2 at point E, was parallel to line DG. On the other hand, F_2 was perpendicular to line DG. So HE was the arm of F_2.

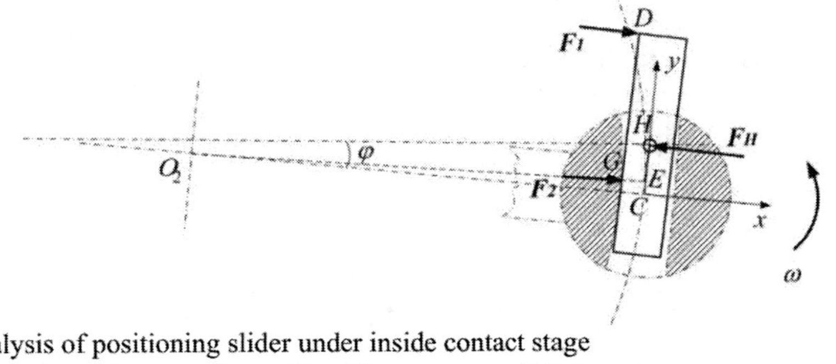

Figure 6. Force analysis of positioning slider under inside contact stage

F_H could be calculated by equations (2), $\sum M_H = 0$ and $\sum F_x = 0$.

$$F_H = \frac{M_C(2R\sin\frac{\varphi}{2} + \sqrt{b^2 + r})}{4R^2\sin^2\frac{\varphi}{2} + 2Rb\sin\frac{\varphi}{2}} \tag{3}$$

(3) Finally taking the driving mechanism as the research object.

Driving torque M_O could be obtained by combining Eq.(3) and equation $\sum M_o = 0$.

$$M_O = \frac{1}{48} C_D \rho h \omega^2 \left((32-16A) R^4 \sin^4 \frac{\varphi}{2} + 24(\mathrm{A}+1) a^2 R^2 \sin^2 \frac{\varphi}{2} + 3Aa^4 \right) \tag{4}$$

where $A = \dfrac{2R\sin\frac{\varphi}{2} + 2b - \sqrt{b^2+r}}{4R\sin\frac{\varphi}{2} + 2b}$ and r was radius of shaft with groove.

4.2.2 The driving torque model under outside contact stage

The driving torque model under outside contact stage could be expressed as:

$$M_O = \frac{1}{48} C_D \rho h \omega^2 \left((32-16B) R^4 \sin^4 \frac{\varphi}{2} + 24(\mathrm{B}+1) \mathrm{a}^2 R^2 \sin^2 \frac{\varphi}{2} + 3B\mathrm{a}^4 \right) \tag{5}$$

where $B = \dfrac{b}{2R\sin\frac{\varphi}{2} + b}$.

5 Simulation analysis of driving torque on driving mechanism

Table 1 . Calculation parameters of driving torque on driving mechanism of HRW

Calculation parameters	Size
Lift coefficient *CD*	4.6
Air density ρ ($\mathrm{kg{\cdot}m^{-3}}$)	1.225
Width of blade h(m)	0.2
Length of blade a (m)	0.141
Length of crank R(m)	0.06
Length of slider $2b$(m)	0.02
Radius of shaft r (m)	0.008

For the convenience of analysis, the initial position of the crank was assumed to be left. Since the angular velocity of crank was 360°/s, drive mechanism could run a circle with 1s. Among them, the stage without local constraint, inside contact stage and outside contact stage were 0.96s, 0.02s and 0.02s. Then Matlab was used to calculated the driving torque's equations (1), (4), (5) with HRW prototype size and related experimental data, as shown in table 1. Finally, we could use Matlab simulated the running process of driving mechanism and the curve of driving torque was obtained, as shown in Fig.7.

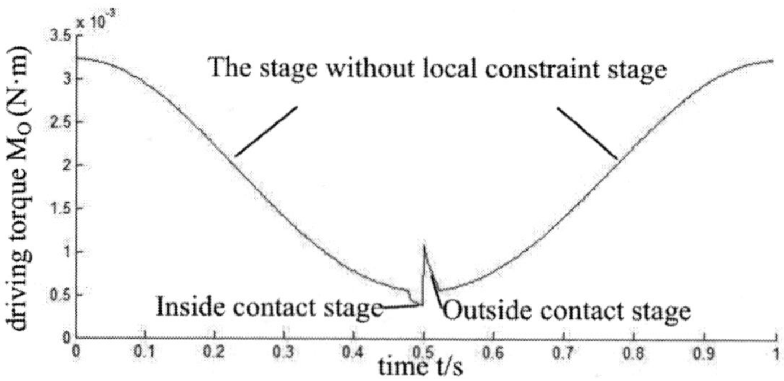

Figure 7 .Driving torque curve on driving mechanism for HRW

Fig.7 showed the driving torque curve of driving mechanism for HRW in one cycle. It could be found that the driving torque uniformly changed over time under the stage without local constraint, but when from the stage without local constraint stage to inside contact stage and the stage from inside contact to outside contact, the driving torque mutation occurred, and the latter was more obvious. The curve could be proved that the local constraint applied on the HRW had influence on driving torque, but not very seriously.

6 Conclusion

(1)Local constraint and its application in the simplified HRM made driving mechanism of HRW more simple in structure and lighter in weight, and the improving of lift-to-weight ratio was more conducive to flying as well.

(2)For the different contact stages of driving mechanism under local constraint, the driving torque model was established. On the basis of model, the curve of the driving torque during operating period were obtained through Matlab, which showed that local constraint had little influence on driving torque. All the work done in this stage lay down the solid foundations for the choosing of Motor and optimization of dynamics.

Acknowledgments

This work is supported by the National Natural Science Foundation of China (Grant No. 51375014 and No. 51405001)

References

1. R.J. Wood, IEEE.Trans.Rob. Autom, **24**,341 (2004)
2. T.N.Pornsin-Sirirak,S.W.Lee.*AICMEMS*,799 (2000)
3. R. Madangopal, Z.A. Khan, S.K. Agrawal, J. Mech. Design,**127**,809,(2005)
4. Z. Wei, L.C.Jia, J.M. Yang. Mech. Eng ,**33**,63,(2011)
5. Z,Z, Qiu, *Press.USTC*,27(2011)
6. X.X Wang, L.L. Mei, F.Q.Chen,Mech.Sci.Tech.Aerosp.Eng, **30**,1020(2011)
7. X.X. Wang, J.H. Qin, H. Qiu, Mech.Sci.Tech.Aerosp.Eng ,**33**,802 (2014)

MATEC Web of Conferences **52**, 02006 (2016)
DOI: 10.1051/matecconf/20165202006

Research on the Application of Tactile Materials to Design and Printing of Books

Wang Yuanyuan [a]

Dalian Polytechnic University, Institute of Clothing, 116034 Dalian Liaoning, China

Abstract. China is a country with the most abundant forms of traditional books, and the evolution of book forms has special material characteristics. The Paper meets the demands of people on multi-sensory stimulation and spiritual experience not only through sense of touch taken as one of human body's multiple senses, but also through analyzing and researching the tactile materials generated based on the sense of touch and discussing how the tactile materials change the single visual experience of traditional paper media. Unique emotional experience and interactive experience brought by the tactile materials are the future book development direction. Through discussing application to contemporary culture and design thinking and introspection and more profoundly understanding the cultural connotation and the artistic value of the books, a newer way of development is provided for future development of paper books.

1 Development of Book Materials

Papermaking and printing invented are the most outstanding cultural heritage in China. As one of ancient forms of art design, book design has a long history and is rich in cultural connotation. In recent years, along with continuously steady and healthy development of economy and culture as well as science and technology, update of book design thinking, change of reader' reading ideas, increasingly-complete printing and publishing system and rapid development of new technologies and new materials, people have been propelled to look for new book forms, therefore, the traditional printing industry presents new development features.

Books are not only one of the important media for people's understanding of nature and understanding of the world, but also are culture's concrete carriers and media[1]. The evolution of Chinese books' forms has unique material characteristics, namely from the earliest inscriptions on oracle bones to bronze inscriptions, inscribed bamboo slips to silk and finally to the reel system and album system of paper materials, different influences and updates of the materials decide the external forms of the books.

The invented printing technology makes the books spread and stored and plays the roles of the most important cultural inheritance and communication. Printed materials are becoming more and more abundant along with continuous improvement of the printing technology. The updated printing technology has also expanded the scope of performance of materials. More types of substrates have offered the possibility in diversified development of book forms.

[a] Corresponding author : 763631431@qq.com

2 Tactile Materials and Printing

The way of reading the books decides that this art form is not confined to the transmission of visual sense. When people pick up a book and gently read it, it is bound to be a behavioral activity closely linked with the tactile materials like a ceremony. Miller Ray once said, "the awakening of the mind comes from the body", which confirms that the suggestion of sensory stimulation to human psychology makes it easier for people to perceive the information transmitted by the books.

As the general term of mechanical stimulations such as contact, sliding, pressure and so on, the sense of touch is a kind of feeling caused as the slight mechanical stimulation makes the skin superficial layer receptor excited. A lot of tactile animals rely on the sense of touch to locate and recognize their living environment and capture their favorite foods. The human beings' tactile organs are all over the body. Taking contact stimulation on human body surface as an example, temperature, sense of pain, sense of touch and other feelings can be felt by means of free nerve endings of human skin. The skin is regarded as the outer layer of the human brain by us, which owns an extremely-sensitive and rich sensory receptor. The skin, which is formed by tissues forming the brain, is the extension of the human brain. The tactile materials have widely applied to book design and printing industry in recent years, which are mainly reflected in two aspects of printed substrates and special printing technology.

2.1 Substrates

As most of traditional substrates are made of paper, along with continuous improvement of papermaking technology in recent years, there are many types of special paper, and paper patterns, textures, softness, density and other different textures can result in a big imagination space so as to bring along inspiration to book designers. For example, there are embossed paper with non-obvious patterns on the surfaces, hot bake paper of which the color is changed through pressing, dermatoglyph paper imitating organisms and animals, wove paper, silkworm cotton paper, velvet paper, abrasive paper with special sense of touch, bamboo paper, pressure sensitive paper with special functions, blueprint paper, thermal-sensitive paper and so on, which can generate materials with different senses of touch and become the new darling of book printing. In the processes of design and printing, although pictures and texts in the books belong to visual perception, each substrate is no longer a simple and traditional two-bit plane. Through touch, the unique physical properties of the substrates and the psychological touches caused in the delicate change can be felt.

Except for paper substrates, a lot of applied non-paper-made materials also provide more choices for the substrates of the books, like leather, special cloth, plastic, wooden boards, metal, silk cloth and so on. In the book Zhu Xi's Qianziwen designed by Lu Jingren serving as a Professor in China Central Academy of Fine Arts, about 1,000 words in the letter were engraved on paulownia wooden boards, and their engraving method imitates wood carving of printing in the Song Dynasty. In the whole letter, all the words were better put together through the belt, which form a book form with unique modeling. He said, "Zhu Xi's Qianziwen was engraved on the stones, which brings along a kind of feeling of cutting by knifes and axes. I hope that people can understand this strength from design and touch its grains".

2.2 Special printing technique

As in today of great impact on paper books, the ordinary printing process has difficultly met people's requirements and imagination on the books, the creation of tactile experience through the special printing process is one of future development directions of the books[2]. Through special technical processing performed on material surfaces by utilizing the modern printing process which is developing rapidly, the materials can better highlight their own characteristics, even the complex characteristics which are highlighted by various materials combined together and are superior to those displayed by single material. For example, stamping process (die stamping) is generally used for depressing and projecting title texts or cover patterns and can achieve texture effects of being rich in sense of touch and highlighting luxury; thermography can create a sense of hierarchy on one plane, even plain paper to give people different feelings and can also skillfully combine double senses, namely visual sense and sense of touch. In addition, there are laser engraving process, UV process, foil stamping process, flocking process, leather carving, die cutting and creasing and other special processes. Qian Cunxun once mentioned in the Chinese Paper and Printing that the continuously-improved process changed the performance of material surfaces and also presented a diversified trend in aspect of selection of book materials[3]. Appeared special paper enriches the texture and the beauty of the books. Applied non-paper materials enrich in the connotation of the books and make the books full of strong breath of times.

When Kenya Hara serving as a Japanese designer designs the program book for the opening ceremony of the Winter Olympics in Nagano, in order to highlight the beauty of ice and snow in the Winter Olympic Games, a kind of special paper, namely "ice and snow paper" was specially designed. On the soft paper as white as snow, the metal template for texts are heated and pressed on the paper through the special processes of denting and thermography, so that the text parts on the paper depress; under the action of heat, paper fibers are melted into translucent states, so that depressed typefaces achieve translucent effects like ice. This special book arouses people's memory of the original winter, reminds people of remembering the clusters of footprints left on the snowfield which is not stepped by others and effectively and artistically conveys the spirit of the winter Olympics[4].

2.3 After-printing-technology

After-printing-technology is the last link of printing, it's an important measure to consummate the printing quality and improve the aesthetic feeling and value of the products. It plays a decisive role in the final form and the applicability of the printings. Many books use after-printing-technology to complete the tactile materials. There is a special technology to deal with the spine, it is called the bare ridge binding. It make the sewing and glue of the book bare, this can make the book spared out completely and appear comfortable smoothly. What is more important, the designers want to show the internal binding structure of the books to the readers and let them touch it. In this way, the readers can comprehend a book through visual and feeling and have a double experience. Besides, there are also many other technology such as cutting technology, can also trigger people's aesthetic feeling. Qu Minyue and Jiang Qian used a special cutting technology in their book *Dance of Dunhuang,* which is the most beautiful book in China 2014, the edge of the kraft paper was specially designed to show the historical broken of the Dunhuang mural. The sound books of children also use the after-printing-technology to add some thin and brittle paper into the mezzanine of these cloth art books. This help to develop children's intelligence through using their eyes, ears and fingers when they touch these hisses books.

The importance and particularity of printed materials are self-evident, and their significances are the results of designer's thinking[5].The advantages of the materials can be fully realized only through the appropriate process. An interactive relationship of mutual interrelation and promotion exists between the materials for printing the books and the process. The perfect process is conductive to embodying the peculiar language idea and feeling of the materials themselves. Therefore, more spaces are expanded for design and development of the books.

In recent years, in order to walk out of the bottleneck and get rid of the shackles caused by traditional forms, although the book design has been changed and broken through in aspects of design style, materials and special printing processes, many works ignored original function and meaning of the books for the purpose of seeking after gorgeous appearance and strong visual impact. Secondly, many books' design means are not better linked with the contents of the books and do not convey accurate themes in the books, so that form is greater than content, and a presumptuous guest usurps the host's role. Materials are the material basis of books, the selection of the materials can not be divorced from the original intention of the books, it has to make it clear that materials are for contents, and the values and significances of the books can be conveyed only by utilizing the material language reasonably and accurately.

3 Multi-sensory Experience of Book Design

Rapid development of times and update of new materials and new technologies drive the change of people's reading ideas. Internet, mobile phones and e-books have become the main ways for young people to read, so the traditional publishing industry and the printing industry are facing a huge impact. Changing the single visual experience caused by traditional paper media, developing the diversified design experience and meeting the demands of people on multi-sensory stimulation and spiritual experience are the future book development direction. Members in the reading experience team in University of Stavanger, Norway think that "reading is the interaction between people and technologies, and as the sense of touch and the physical properties of the printed books generate different cognitions and emotional experiences for people, it is especially important for people to focus on reading".

Organs for feeling outside things by human beings include eyes, mouth and nose, ears, skin and so on, namely stimulations caused by visual sense, sense of touch, sense of hearing, sense of smell and sense of taste as well as a variety of collections from these senses are integrated in the brain to finally form an image. When we read a book, the appearance, the weight, the color, the paper, the patterns and even the smell of ink emitted from books wholly form our understanding of the books. This is called unique interaction and experience transfer among all sensory organs (also known as the "synaesthesia"). The synaesthesia requires that the modern book design can stimulate the readers' sensual experience from all angles.

From another point of view, the innovative development of printing technology and process make full material preparations for design and printing of the modern books. As people's aesthetic requirements for the traditional books are getting higher and higher, the readers need to get a synesthesia experience of beauty from five senses. Taking children's books as an example, the traditional books take "paper" as a medium, and the children's books are difficultly confined to single materials, and their readers are eager to get more material touches from the books. Through study, psychologists have found out that the development of sense of touch is very important for children's psychological development. Some children's books are made of some cotton cloth or flannelette, thereby being soft, safe and environment-friendly and dust-proof. Toy books made of safe plastics can be prevented from being tore by children and be waterproof and cleanable cleaned, so that the children can take baths while reading. In addition, there are some natural books capable of training the children's sense of touch in a targeted manner, like "Journey of Nature" compiled by Rongxin Culture in which various special materials which imitate animal furs in nature are placed in its beautiful figures, so that while reading the books, the children like touching the wonderful natural world personally. The multiple sensitive explorations of sense of touch are conductive to children's action and cognitive development. The sense of touch is the basis of assisting other sensory systems to achieve functions. Based on the touch type books, the children can touch different materials in combination with the figures and recognize and identity that objects are smooth, soft, coarse and sharp. Through various stimulations, the children's perception of all kinds of things can be enhanced. American Dr. Ayres proposed the concept of sensory integration in 1972 and thought that the sensory integration is the ability that the human body effectively utilizes own sense organs in the environment to obtain and input information of different sensory pathways from the environment to the brain, and then the brain analyzes the inputted information for many times, performs overall treatment and makes adaptive response. Through sensory integration training, not only is the comprehensive development of children's visual sense, sense of hearing, sense of smell, sense of taste, sense of touch, vestibular sense and proprioceptive sense stimulated, but also the children's comprehensive integration ability is promoted.

The sense of touch is not the most important function of aesthetic appreciation, but is inextricably linked with visual aesthetic appreciation. The application of the tactile materials to book design makes people more strongly feel different skin textures and hardness of the books, even temperature and other differences in different aspects, which can bring along unique spiritual feeling to people[6]. Each material has its own expression and can express its own emotion and give people different psychological suggestions. For example, the grain materials have a warm feeling, the metal materials have a cold feeling, and the wood materials give people a natural and comfortable feeling. As material carriers having the communication functions and serving as artistic media, the materials self contain rich culture, and the culture is the important link to reflect the designers' ideas and convert book contents into materialized book forms, and the stress on the use of material thinking is a new design development direction of the modern books. Arnheim Rudolf says when taking about the sense of touch that art is from the body sense experience at any time and at any place[7]. The tactile experience is a tactile feeling of effectively stimulating and reflecting audiences participating in the design. The feeling caused by sense of touch is more subtle and delicate than that caused by visual sense and more can touch people's hearts.

4 Future and Development of Tactile Materials in Book Printing

Although China is a country with the most abundant forms of traditional books, it is short of new ideas in the design of modern books. Foreign excellent publishing houses and book designers have designed pop-up books, audio books, touch books, lift-the-flap books, interactive books and other multiple innovative books in succession. However, China is providing the services of printing in China and port-printing process at present only.

In August, 2003, the Shanghai Publishing and Press Bureau connected with the German Book and Art Foundation, decided to hold a selection called "the most beautiful book in China". This selection has been held 13 times so far, it had selected hundreds of books and helped them entered the "the most beautiful book in the world" selection. This is an important step of our country to change our character from printing country to designing country, with the help of the book competition in Leipzig, German, show our well-designed books to the world.

Lv Jing is a famous book artist in China. He is always paying his attention in the processing of the development of book art and founded the first professional platform in China to deal with process testing, commercial marketing, art exhibiting and academic exchanging.

He thinks that the design idea of the book design needs to be performed in combination with design, process and materials, the design only completes a small part of the whole book, and more attentions to design principle, artistic aesthetic feeling and cultural taste of the books must be paid[8]. We not only need to understand the properties of the materials and are proficient in the process application, but also dare to innovate and challenge the process. As special commodities, the books not only are used as text carriers to transmit the information, but also carry a nation's history, culture, folk life and art design. While facing a test, China's modern book design and printing industry adapt to new needs and seek for new development. Five-sense synesthesia-dominated design experience becomes the future book development trend. As the sense of touch is one of the important forms of human perception, if the tactile materials are blended into book design and printing, readers not only can passively accept during reading, but also experience the fun of interactive experience and are endowed with more special reading experiences.

This kind of touch experience is not a mordern touch screen technology, this new type of human-computer interaction has replaced the traditional button panel, it can cmplete the operation and creat vivid video effects more simply,directly and naturally. But it still can't get rid of the fate of dealing with simple electronic products. All kinds of electronic products in modern life make people difficultly read deeply, and too much senses of science and technology and functional pursuit make people almost forget the fun of reading itself.

The fundamentality of design is to be people oriented and meet people's aesthetic needs. It is expected that there are more outstanding book designers in China in future, and more beautiful books enter Leipzig. It is expected that more readers realize the importance of emotion in book design and have a more profound understanding of cultural connotation and artistic value of the books, and a more new way of development is provided for future development of paper books.

References

1. X.J .Chen,Peking university press, *paper* (2012)
2. Y.D .Hu, Y.F. Zhang, Printing industry press ,Paper to print professional(2014)
3. C.X .Qian, Guangxi Normal University Press, *Chinese paper and printing A Cultural History* (2004)
4. Harakenya, Shandong people's publishing house, Design of design (2007)
5. J.P. Hao, Shanxi Normal University, *Modern books design material of language study* (2012)
6. W. Ding, Shanghai normal university, *To explore the tactile design of books* (2012)
7. Rudolf Arnheim，sichuan people's publishing house，*visual thinking* (1998)
8. J.R.Lv, China youth press, *Book design* (2009)

MATEC Web of Conferences **52**, 02007 (2016)
DOI: 10.1051/matecconf/20165202007

Adaptive Robotic Systems Design in University of Applied Sciences

Jos Gunsing, Fons Gijselhart, Nyke Hagemans, Hans Jonkers, Eric Kivits. Peter Klijn, Bart Kapteijns, Diederich Kroeske, Hans Langen, Bart Oerlemans, Jan Oostindieand Joost van Stuijvenberg

Avans University of Applied Sciences, Centre of Expertise for Sustainable Innovation, Research Group for Robotics & Mechatronics, Breda/'s-Hertogenbosch, The Netherlands

Abstract. In the industry for highly specialized machine building (small series with high variety and high complexity) and in healthcare a demand for adaptive robotics is rapidly coming up. Technically skilled people are not always available in sufficient numbers. A lot of know how with respect to the required technologies is available but successful adaptive robotic system designs are still rare. In our research at the university of applied sciences we incorporate new available technologies in our education courses by way of research projects; in these projects students will investigate the application possibilities of new technologies together with companies and teachers. Thus we are able to transfer knowledge to the students including an innovation oriented attitude and skills. Last years we developed several industrial binpicking applications for logistics and machining-factories with different types of 3D vision. Also force feedback gripping has been developed including slip sensing. Especially for healthcare robotics we developed a so-called twisted wire actuator, which is very compact in combination with an underactuated gripper, manufactured in one piece in polyurethane. We work both on modeling and testing the functions of these designs but we work also on complete demonstrator systems. Since the amount of disciplines involved in complex product and machine design increases rapidly we pay a lot of attention with respect to systems engineering methods. Apart from the classical engineering disciplines like mechanical, electrical, software and mechatronics engineering, especially for adaptive robotics more and more disciplines like industrial product design, communication &multimedia design and of course physics and even art are to be involved depending on the specific application to be designed. Design tools like V-model, agile/scrum and design-approaches to obtain the best set of requirements are being implemented in the engineering studies from the early beginning.

1 Introduction

Avans University of Applied Sciences is situated in an industrialized part of the Netherlands.

In the technical field a shortage of highly skilled people is present. Thus production automation and robotics is an important theme. Many small and mediumsized companies are active in this industry notwith mass production but with building highly complex machines in small series. Also logistics/transport and material handling is important. Increasing logistic flexibility and a huge variety of products to be handled are typical trends. Apart from this in many types of industry a shortage of highly skilled technical people is present. Thus automation and robotics are important themes in combination with flexibility in capacity and the type of robot actions. With the industry in our region we agreed on focussing on robot gripper technology in combination with adaptive robotics.

Being a university of applied science we concentrate not so much on research into new technologies but we aim for making these new technologies applicable in new innovative products and systems. We do so in cooperation with universities and companies. Most of the projects we carry out in closecooperation with other universities of applied sciences like Fontys, Utrecht, Saxion and NHL. Most of the projects are funded by the universities of applied sciences themselves in combination withprojects funding (governmental organisation SIA ;Innovation Alliance Foundation)

The choice of the projects is the outcome of a process in which we take into account the business and product roadmaps of the involved companies and we combine these insights with very useful technology roadmap plus research agenda information from the organisation euRobotics[1, 2].

In the projects the research is being carried out with teachers, students and companies working together and all of them will increase their knowledge. Our students are our main workforces. The teachers are coaching them in the research process and go deep in the complex matters themselves.With this knowledge education and product innovation will be lifted to a higher level. Due to the increased system complexity and the increased multidisciplinarity we put also a lot of effort in education in systems engineering. We do aim for systems engineering as an umbrella under which many systematic design approaches can be applied like V-model, scrum/agile etc.

2 Adaptive Robotic Systems Design

2.1 Systems setup

Having chosen for gripper design as a focus subject we had to decide on which applications we should work for; In our case we choose for focus on industrial systems. Nevertheless we have had a chance to work on healthcare applications as well. We work on those types of healthcare gripper technologies, which are also applicable in industry for handling purposes.In the project roadmap for our research we aim for gradual increase in robot complexity especially in terms of robot/human cooperation. Based on a systematic approach in requirement analysis, choice of functions and modular design we have been working on several applications. More recently we have started with involvement of communication & multimediadesign people in robot/human interaction. It is our strong opinion that they can help is to define the right functionality and to design a robot with which people are really happy. Also involvement of artists can help to magnify or even exaggerate robot functional aspects such that it helps people to come up with the right robot design ideas.

2.1.1 Industrial applications

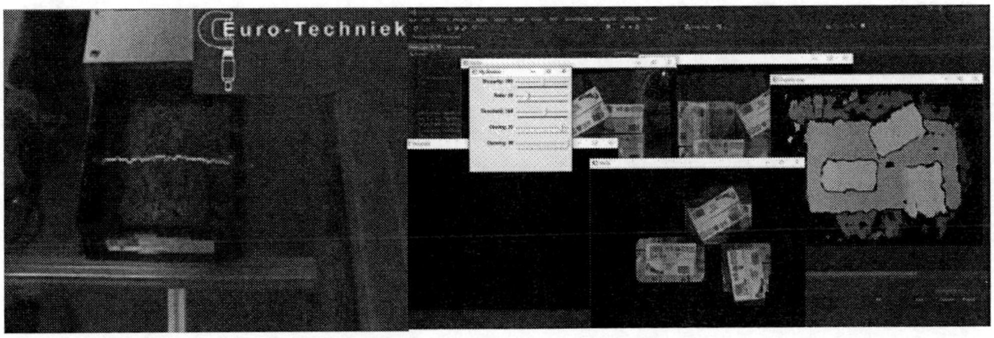

Figure 1. Binpicking application; scanningof non-oriented components in the bin. Box recognition and handling application: disparity map generation for box recognition; only upper layer of boxes should be visible.

At this moment we have been working on binpicking applications for several years already. In our region many companies do work with metal or plastic components. Picking out the unsorted and non-oriented components out of a binis a challenging job especially when you are aiming for a very short learning time for the robot/vision/gripper system when trying to handle a new component. We are aiming now for learning times of one hour instead of weeks to obtain a robustly working system.

In a project in the near future we will be working on a robot-buddy for mechatronic systems assembly engineers. This engineer will teach his robot-buddy himself in order to take over part of the more routine-like jobs like putting in/screwing in bolts/nuts etcetera. Because he teaches this robot-buddy himself the buddy becomes more or less part of himself.

2.1.2 Healthcare applications

We are involved in a healthcare robotics project in which we develop a healthcare robot which is meant to help elderly people picking up objects which are on the floor or have fallen.We concentrated on the hand gripper function. Healthcare technology students studied the hand functions very thoroughly and came up with a set of user requirements includinguse cases, which are the basis for further development and engineering. During this process the healthcare technology department is still involved when reviewing the gripper design and especially when testing the gripper and robot system with the target user group.

2.2 Technology focus

In 2.1 we described the applications of the grippers we design. Several technologies will have to be a permanent subject of study. We apply several systems engineering design approaches over the whole process from user requirement/use case set up to concept generation/modelling, design , engineering and experimental validation including literature study in all those stages of the design process. We consider this to be the backbone of our educational program for esp. the mechatronics engineers but we aim to involve also other disciplines in the near future. Companies in our region struggle with the implementation of systems engineering but they are very much aware that it will help them to work successfully in multidisciplinary teams, to work together with other companies, to design modular product families, to change from a engineering-to-order to a configure-to-ordercompany.

Amongst others we base ourselves on the CAFCR-method [3], (Customer, Application, Functional, Concept, Realization views with respect to the product or system which must be developed) for obtaining the most suitable requirement/product concept combination.In gripper design you will find the complete set of mechatronic technologies you find in a robot (In the next chapter we will go into more detail with respect to these technologies):

- gripper design
- force and slip sensing plus control
- actuation/transmission
- 3D vision incl. data transfer At this moment we are gradually combining all the developed technologies in a complete gripper and robot system together with our cooperation partners in industry and (applied) sciences

3 Technology

3.1 Gripping

3.1.1 Gripper system

We have considered many alternatives as a gripper concept. It had to fulfil at least following requirements (in this case a healthcare robot for picking items from the floor):

- Two finger gripper are sufficient for the required jobs/use cases
- Underactuated gripper design for cases of simplicity
- Easy manufacture/low product cost/easy exchangeability

The design, which proved to be the best proposition, has following features.

- All moving parts manufactured in one piece; waterjet cutting + milling from polyurethane;
- Simple design with use of elastic hinges with relatively low hysteresis
- External actuation possible; in this case with twisted cord plus electrical motor
- Moving parts covered with 3D printed covers

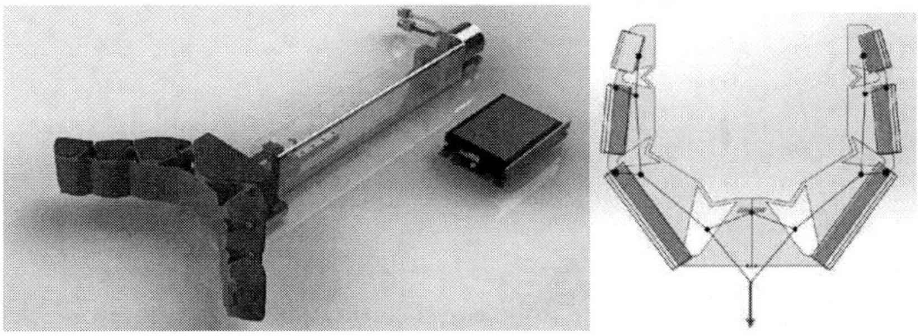

Figure 2. Complete gripper plus cross-sectional view of the underactuated gripper design with elastic hinges

At this moment only a limited amount of testwork has been carried out. First testresults are promising but extensive testwork including validation tests with potential users must be carried out.

3.1.2 Force and slipsensing/-control

Some time ago we started work on measuring and controlling the force in the gripper and we found out that it would be most appropriate also to measure and control slip when holding objects between the fingers. We use a slip-sensingconcept developed by the UEC Shimojo Laboratory (University of Tokyo). This slip sensor is based on a conductive rubber placed on top of an electrode. In pressurized condition the conductivity changes and a force-dependent signal can be measured. In [4] it is stated that just before slip occurs a high frequent signal occurs. At this moment we have a test rig up and running and the first promising indication of slip sensing i.e. high frequency signals have been seen. The test rig set up has been designed for very low hysteresis in the gripper finger force in both holding and slipping direction.

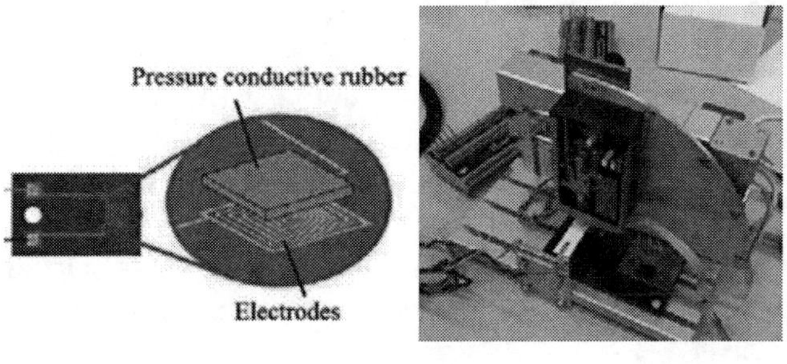

Figure 3. Force and slip sensitive sensor with pressure conductive rubber [4] andtestrig

Also a test gripper with very low friction/hysteresis has been designed. After evaluation of the results on this test rig we will be able to put the force and slip sensor on the gripper which was described in 3.1.1., which is much more robust but also has a higher hysteresis due to the polyurethane material choice.

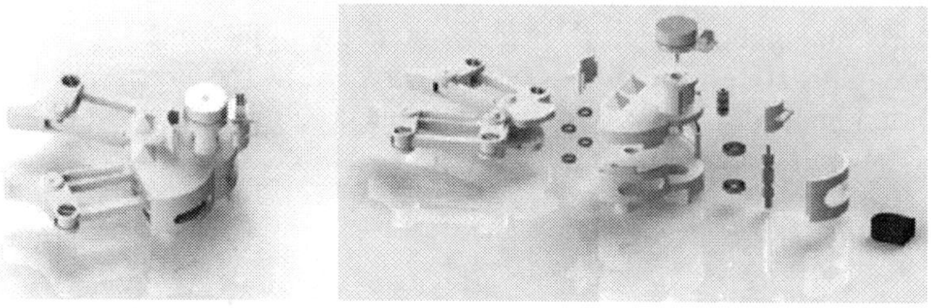

Figure 4. Low hysteresis gripper design for laboratory test; development and validation of force and slip sensor

3.2 Actuation/transmission

Our choice for a combination of a twisted cord and electrical motor has not been based on extensive studies but it appeared to be a actuator/transmission which was not quite known yet and papers were seldom at that time. A twisted cord consists of 2 or more cords twisted together; one end is fixed; the other end is directly attached to a motor shaft; when the motor turns the total length will become shorter and it can exercise a longitudinal force. This force can be applied to one or more gripper fingers.This actuator has its advantages when using it for moving the fingers in a gripper. The electrical motor can be placed in the arm of a robot at some distance keeping the gripper compact and the inertia low when moving the gripper together with the arm. A high torque motor is not necessary to move the fingers. In order to become more acquainted with the characteristics of a twisted cord we did set up a theoretical model and we also designed and built a test rig in order to predict and measure the relation between the force/torque and shortening of the twisted cord/number of motor revolutions.

Figure 5. Twisted cord actuator/transmission; performance testrig, twisted cord and twisted cord model.

3.3 Vision incl. data handling

No robot and gripper system without a vision system ; since the early start of our robot investigations we have built up experience with vision systems based on several types of sensors;

- laser triangulation sensor
- time of flight camera
- stereovision(2 camera's)

We built up experience with these 3 types of vision systems; Also we have put a lot of effort in setting up our own FPGA based data handling system. This has been a big job and although we have obtained a lot of inside knowledge in this system we will now change to a purchased system and we will concentrate on the stereovisionsystems integration with the robot and gripper system including actuation force and slip sensing system. Data handling is also important; since robots tend to work (partially) autonomous we have several possibilities:

- data transformation to information is carried outdirectly on the robot preferably close to the camera
- data being sent over by Wi-Fi to operator dashboard or data storage depending on the application

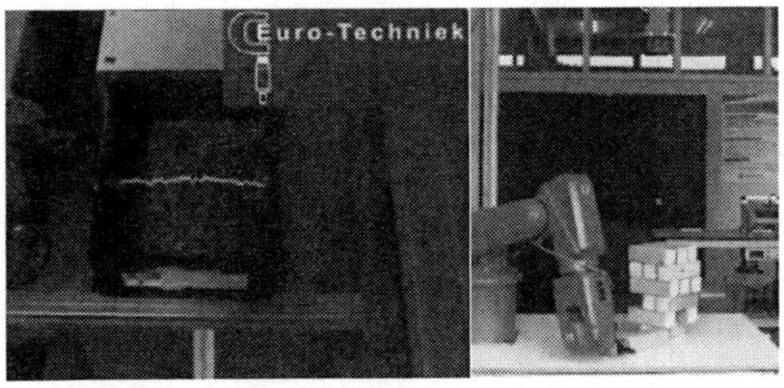

Figure 6.3D (2,5D) vision systems; laser triangulation and time of flight

4 Conclusions & Recommendations

4.1. Technical

In about a year from now we will have working gripper & robot systems available. Meanwhile we will continue our research into increased levels of adaptive cooperative behaviour and of robots. Application of knowhow from artificial intelligence, behavioural sciences and communication &multimedia design will help to further expand adaptive robotics applications.At this moment projects with these disciplines together with engineering are being set up as an eye opener for all those interested in robots and humans working more closely together.

4.2. Educational

Working with students in applied research helps them to develop an innovation-minded attitude while being very critical about their own way of working. Systems Engineering helps to line up process thinking and helps with tools for further development of their skills and mind.Meanwhile we incorporated research into the educational programs of the years3 and 4 (4 year educational track)

References

1. Strategic Research Agenda For Robotics in Europe 2014-2020, SPARC Partnership for Robotics in Europe
2. Robotics 2020 Multi-Annual Roadmap for Robotics in Europe, Call 1 ICT23-Horizon 2020, Initial Release B 15/01/2014
3. G. Muller, A Multi View Method for Embedded Systems Architecting, Balancing Genericity and Specificity, thesis 2004, ISBN 90-5639-120-2, http://www.gaudisite.nl
4. S. Teshigawara,T. Tsutsumi,S. Shimizu, Y. Suzuki, A. Ming,M. Ishikawa, M. Shimojo, Highly Sensitive Sensor for Detection of Initial Slip and Its Application in a Multi-fingered Robot Hand, 2011 IEEE International Conference on Robotics and Automation pp.1097-102, 2011.

MATEC Web of Conferences **52**, 03001 (2016)
DOI: 10.1051/matecconf/20165203001

A Tooth Flank Crowning Method by Applying A Novel Crossed Angle Function Between the Hob Cutter and Work Gear in The Gear Hobbing Process

Yu-Ren Wu[1,a], Van-The Tran[2] and Ruei-Hung Hsu[3]

[1]*Department of Mechanical Engineering, National Central University, 300, Jhongda Rd., Jungli City, Taoyuan 320, Taiwan*

[2]*Department of Mechanical Engineering, Hung Yen University of Technology and Education, 5 Rd., Hung Yen City, Vietnam*

[3]*Bachelor's Program in Precision System Design, Feng Chia University, 100 Wenhwa Rd., Seatwen, Taichung 40724, Taiwan*

Abstract. In this paper, a novel longitudinal tooth flank crowning method is proposed by setting the crossed angle between the hob cutter and work gear as a linear function of hob's traverse feed movement in the gear hobbing process. However, this method makes twisted tooth flanks on the hobbed work gear. Therefore, a variable pressure angle hob cutter is applied to obtain an anti-twist tooth flank of hobbed work gear. A computer simulation example is performed to verify the superiority of the proposed novel hobbing method by comparing topographies of the crowned work gear surfaces hobbed by a standard hob cutter and a variable pressure angle hob cutter.

1 Introduction

In the conventional gear hobbing process, the longitudinal crowning for tooth flank on a helical gear surface is usually accomplished by varying the center distance between the hob cutter and work gear, and treated the crossed angle of the hob cutter and work gear as a fixed machine-tool setting. However, in this study, the center distance between the hob cutter and work gear is fixed and the crossed angle is set as a linear function of hob's traverse feed movement for longitudinal crowning tooth flank. Thereby it will induce a twist of tooth flank on the hobbed work that is the reason for a bias of tooth wear and noise of the meshing gear pair. Therefore, to obtain an anti-twist tooth flank of the work gear in longitudinal tooth crowning, a variable pressure angle (VPA) hob cutter is applied during gear hobbing process.

Several researches about hobbing method are recently presented for longitudinal crowning tooth flank of helical gear surfaces. Firstly, a method for longitudinal crowning tooth flank of involute helical gear by varying the center distance was proposed by Litvin et al. [1]. Subsequently, Wang and Fong [2] was proposed a novel face-hobbing method to generate spur gears with longitudinal crowning by two head cutters form an imaginary generating rack with cycloidal lengthwise tooth traces. After that, a mathematical model with a two-parameter equation of meshing for simulation of

[a]Corresponding author : yurenwu@ncu.edu.tw

gear hobbing process was proposed by Chen et al. [3]. Previously, Witte [4] proposed a twist free hobbing solution by analyzing the width of convexity and amount of bias of the tooth flank as well as a finishing hob to cut external crowned helical gears without profile bias. After that, Lange [5] presented some important discussions on the tooth flank's bias errors of helical gears caused in conventional cutting methods and offered techniques used to limit and eliminate bias errors. In the same time, a method for varying the hob profile angle along its lead and shifting the hob diagonally over the whole tooth face width during gear hobbing to reduce the twist of work gear tooth flanks was patented by Winkel [6]. Recently, Hsu and Fong [7] patented the design of a variable tooth thickness (VTT) hob cutter in it's axis and the method of hobbing with a diagonal feed and without varying the center distance. More recently, Tran et al. [8] proposed a novel methodology for finish-hobbing of the twist-free longitudinal crowning tooth flank of helical gears by using a new the hob's diagonal feed motion and a dual-lead hob cutter. And a hobbing method to attain anti-twist tooth flanks of the helical gear in longitudinal tooth crowning by supplementing an additional rotation angle of work gear during its hobbing process is presented by Tran et al. [9]. Finally, the overall geometry, design, and manufacture of a conventional helical gear were detailed in textbook by Litvin and Fuentes [10].

In this research, a mathematical model for anti-twist tooth flank of the work gear surface, hobbed with a new crossed angle and a VPA hob cutter, is constructed in longitudinal crowning process. A numeral example is also presented to illustrate the merits of the proposed gear hobbing method.

2 Mathematical model of the VPA hob cutter

The profile of the generating surface of the standard hob cutter is the same as that of the standard helical gear that is generated by a standard rack cutter. To obtain an anti-twist tooth flank of helical gear, the tooth profile of the standard rack cutter is modified with pressure angle changed in it's longitudinal direction, as shown in Fig. 1(a). The generation coordinate systems of the VPA hob cutter are shown in Fig. 1(b), where the coordinate systems S_7, S_3, and S_4 are rigidly connected to the rack cutter, hob cutter, and frame, respectively. According to Fig. 1, the position vector and unit normal vector of the rack cutter's right-hand side profile can be expressed in coordinate system S_3 as follows:

$$\mathbf{r}_3 = \mathbf{M}_{37}\cdot\mathbf{r}_7 = \begin{bmatrix} \cos\psi_1 & -\cos\beta_{o1}\sin\psi_1 & -\sin\beta_{o1}\sin\psi_1 & r_{o1}(\cos\psi_1+\psi_1\sin\psi_1) \\ \sin\psi_1 & \cos\beta_{o1}\cos\psi_1 & -\sin\beta_{o1}\cos\psi_1 & r_{o1}(\sin\psi_1+\psi_1\cos\psi_1) \\ 0 & -\sin\beta_{o1} & \cos\beta_{o1} & 0 \\ 0 & 0 & 0 & 1 \end{bmatrix}\cdot\begin{bmatrix} u_1\cos\alpha_{on} \\ -u_1\sin\alpha_{on}+bu_1v_1 \\ v_1 \\ 1 \end{bmatrix}, \quad (1)$$

and

$$\mathbf{n}_3 = \mathbf{L}_{37}\cdot\mathbf{n}_7 = \begin{bmatrix} \cos\psi_1 & -\cos\beta_{o1}\sin\psi_1 & -\sin\beta_{o1}\sin\psi_1 \\ \sin\psi_1 & \cos\beta_{o1}\cos\psi_1 & -\sin\beta_{o1}\cos\psi_1 \\ 0 & -\sin\beta_{o1} & \cos\beta_{o1} \end{bmatrix}\cdot\begin{bmatrix} -bv_1+\sin\alpha_{on} \\ \cos\alpha_{on} \\ -bu_1\cos\alpha_{on} \end{bmatrix}. \quad (2)$$

where u_1 and v_1 are the surface parameters, α_{on} is the pressure angle of the rack cutter, and b is VPA coefficient of rack cutter, ψ_1 is the generated rotation angle of hob cutter around z_3 -axis, and $s_{on} = r_{o1}\psi_1$ is the rack cutter translates along z_4 -axis and r_{o1} is the radius of the operating circle of hob cutter, as shown in Fig. 1.

According to the theory of gearing, the equation of meshing between the right-hand side rack cutter surface and left-hand side hob cutter surface can be obtained by

$$f_1(u_1,v_1,\psi_1) = \mathbf{n}_3\cdot\frac{\partial[x_3(u_1,v_1,\psi_1), y_3(u_1,v_1,\psi_1), z_3(u_1,v_1,\psi_1)]}{\partial\psi_1} = 0. \quad (3)$$

The tooth profile and its unit normal of the VPA hob cutter are defined by considering Eqs. (1), (2) and (3), simultaneously. The locus tooth profile and its unit normal of the standard hob cutter are determined by setting VPA coefficient, b, equal zero.

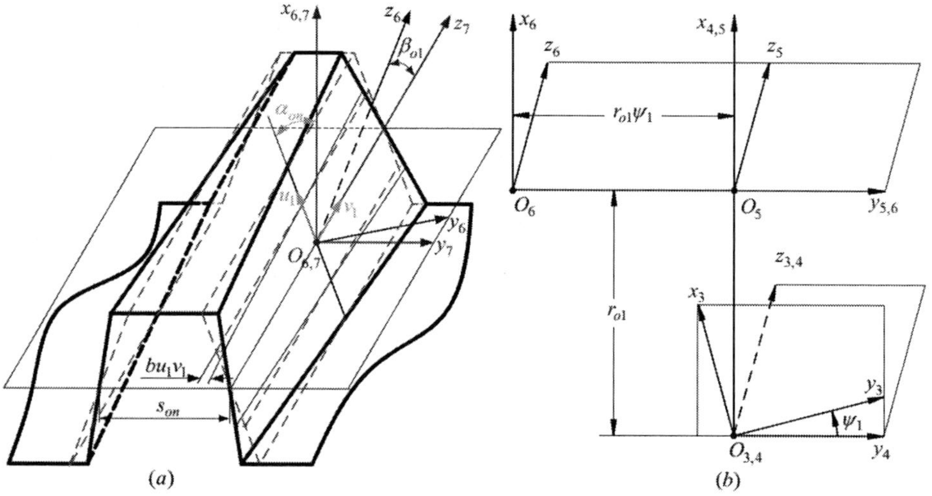

Figure 1. Surface parameters of VPA rack cutter and generation coordinate systems of VPA hob cutter.

3 Mathematical model of the longitudinal crowned work gear on gear hobbing machine

The Coordinate systems for a gear hobbing machine are shown in Fig. 2. Wherein coordinate systems S_1, S_2 and S_a are rigidly connected to the hob cutter, work gear and frame of hobbing machine, respectively. While S_b, S_c, S_d, S_e and S_f are auxiliary coordinate systems for simplification of coordinate transformation. In gear hobbing process, there are two hob's movements: traverse feed movement along the axis of the work gear $z_a(t)$, and diagonal feed along the axis of hob cutter $z_s(t) = a.z_a(t)$. And the crossed angle γ_o of the hob cutter and work gear is usually fixed as a machine-tool and it can be expressed as follows

$$\gamma_o = \beta_{o1} \pm \beta_{o2}, \tag{4}$$

where the β_{o1} and β_{o2} are helix angle of hob cutter and work gear, respectively. The "$\pm$" sign indicates the same and opposite direction, respectively, of helices between the hob cutter and work gear. for longitudinal crowning tooth flank of work gear, the crossed angle (designated as γ) is modified as a linear function of hob's traverse feed movement in the gear hobbing process as follows:

$$\gamma = \gamma_o + c \cdot z_a(t). \tag{5}$$

where c is crossed angle coefficient between the hob cutter and work gear.
According to Fig. 2, by applying the homogeneous coordinate transformation matrix from coordinate system S_3 to S_2, the locus of hob cutter can be represented in coordinate system S_2 as follows:

$$\mathbf{r}_2(u_1, v_1, \psi_1, \phi_1, z_a(t)) = \mathbf{M}_{23}(\phi_1, z_a(t)) \cdot \mathbf{r}_3(u_1, v_1, \psi_1). \tag{6}$$

The rotating angle of work gear $\phi_2(\phi_1, z_a(t))$ is set as a linear function of both the hob's rotation angle ϕ_1 and the amount of hob's traverse feed $z_a(t)$. This rotational relationship between the hob and work gear is defined as

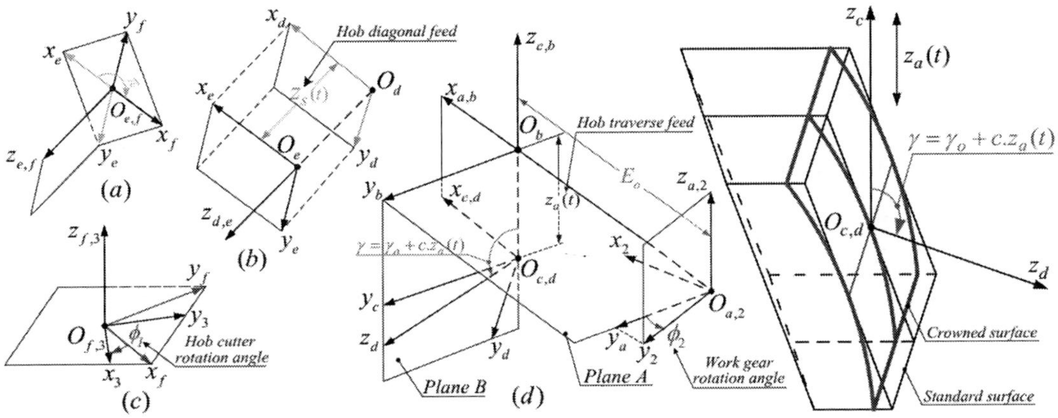

Figure 2. Coordinate systems for a gear hobbing machine.

$$\phi_2(\phi_1, z_a(t)) = \frac{N_1}{N_2}\phi_1 + \frac{\tan\beta_{o2} + a\sec\beta_{o2}\sin\beta_{o1}}{r_{o2}} \cdot z_a(t), \tag{7}$$

where symbols N_1 and N_2 indicate the number of teeth of the hob cutter and gear, respectively, r_{o2} is the radius of the operating circle of work gear, and a is the hob diagonal shift coefficient.

Since the gear hobbing process has two independent motion parameters, ϕ_1 and $z_a(t)$, there are two equations of meshing between the hob and work gear that can be obtained and expressed as follows:

$$f_2 = \mathbf{n}_2 \cdot \frac{d\phi_1}{dt}\left(\frac{\partial\left(x_2(u_1,v_1,\psi_1,\phi_1,z_a(t)), y_2(u_1,v_1,\psi_1,\phi_1,z_a(t)), z_2(u_1,v_1,\psi_1,\phi_1,z_a(t))\right)}{\partial\phi_1}\right) = 0, \tag{8}$$

and $$f_3 = \mathbf{n}_2 \cdot \frac{dz_a(t)}{dt}\left(\frac{\partial\left(x_2(u_1,v_1,\psi_1,\phi_1,z_a(t)), y_2(u_1,v_1,\psi_1,\phi_1,z_a(t)), z_2(u_1,v_1,\psi_1,\phi_1,z_a(t))\right)}{\partial z_a(t)}\right) = 0, \tag{9}$$

where

$$\mathbf{n}_2(u_1,v_1,\psi_1,\phi_1,z_a(t)) = \mathbf{L}_{23}(\phi_1,z_a(t)) \cdot \left[n_{x_3}(u_1,v_1,\psi_1), n_{y_3}(u_1,v_1,\psi_1), n_{z_3}(u_1,v_1,\psi_1)\right]^T. \tag{10}$$

and $\mathbf{L}_{23}(\phi_1, z_a(t))$ is the submatrix of the transformation matrix $\mathbf{M}_{23}(\phi_1, z_a(t))$ by deleting the last column and row.

The tooth surface of the crowned helical gear can be determined by solving Eqs. (3), (6), (8), (9) and (10), simultaneously.

4 Numeral example

The purpose of this example is to validate the proposed longitudinal crowning method. The basic data for work gear, hob cutter and hobbing process are given in Table 1. The level of tooth flank twist of the two double-crowned work gear surfaces is indicated by using two indices: maximum tooth flank twist (designated M_{ce}) defined by the largest crowning amount subtract the smallest crowning amount and longitudinal crowning evenness ratio (designated R_{ce}) defined by the smallest crowning amount divided by the largest crowning amount at different position on the edges of work gear surfaces.

Table 1. Basic data of the gear and hob cutter

Work gear data			
Number of teeth (N_2)		50	
Normal module (m_{pn})		3	
Normal circular-tooth thickness (s_{pn2})		4.71 mm	
Normal pressure angle (α_{pn})		20°	
Helix angle (β_{p2})		20° R.H.	
Face width (F_{w2})		18 mm	
Operating center distance (E_o)		120.51 mm	
Hob cutter data		***Hobbing process data***	
Number of teeth (N_1)	1	Original operating crossed angle (γ_o)	107.89°
Helix angle (β_{p1})	87.89° R.H.	Hob diagonal shift coefficient (a)	-3.26
Normal circular-tooth thickness (s_{pn1})	4.71 mm	Crossed angle coefficient (c)	-1.50×10^{-3}

The normal deviations and tooth surface topographies of the crowned helical gear, generated by applying the standard and VPA hob cutters, are calculated and shown in Table 2 and Figs. 3-4, respectively. The maximum tooth flank twist for the crowned work gear tooth flank generated by using the standard hob cutter $(b=0)$, $M_{ce}=19.56\mu m$, is much higher than that of the crowned work gear tooth flank generated by using the VPA hob cutter $(b=-5.98\times10^{-6})$, $M_{ce}=0.87\mu m$. Besides, the crowning evenness ratio for the tooth flank surface generated by using the standard hob cutter, as shown in Fig. 3 $(R_{ce}=0.19)$, is much smaller than that of the tooth flank surface generated by using the VPA hob cutter, as shown Fig. 4 $(R_{ce}=0.94)$. It is verified that the twist of the tooth flank almost equals zero by applying the proposed VPA hob cutter.

Table 2. Topography characteristics of tooth flank on the crowned gear

Items	VPA coefficient (b)	Maximum tooth flank twist ($M_{ce}(\mu m)$)	Crowning evenness ratio (R_{ce})
Standard hob cutter	0	19.56	0.19
VPA hob cutter	-5.98×10^{-6}	0.87	0.94

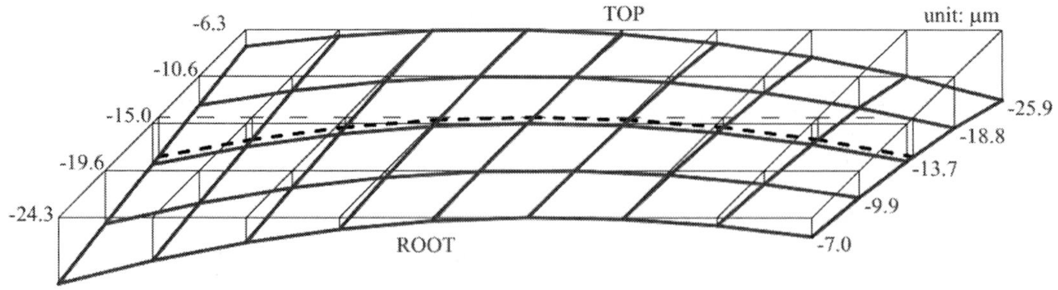

Figure 3. Topography of tooth flank on work gear surfaces hobbed by the standard hob cutter.

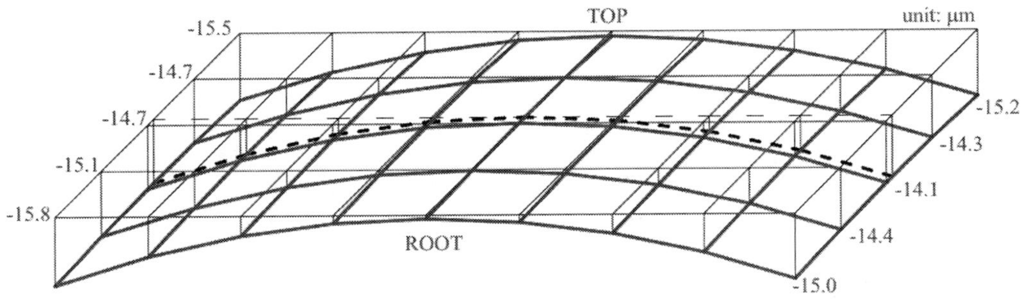

Figure 4. Topography of tooth flank on work gear surfaces hobbed by the VPA hob cutter.

5 Conclusions

In this paper, a mathematical model for tooth flank of the crowned work gear surface is constructed in gear hobbing process. According to the simulated results of numeral example, it reveals that the tooth flank of helical gear can be crowned longitudinally by setting the crossed angle between the hob cutter and work gear as a linear function of hob's traverse feed movement and using a standard hob cutter. However, the using of standard hob cutter will induce twist of the tooth flank on the hobbed gear. Therefore, a VPA hob cutter is applied to obtain anti-twist tooth flank. Thereby the twist of tooth flank on crowned work gear is reduced significantly.

Acknowledgments

The authors are grateful to the Ministry of Science and Technology of the R.O.C for the financial support, contract No: MOST 104-2221-E-035-087- and MOST 104-2221-E-008 -007-.

References

1. F.L. Litvin, A. Fuentes, I.G. Peres, L. Carvenali, K. Kawasaki, R.F. Handschuh, Modified Involute Helical Gears: Computerized Design, Simulation of Meshing and Stress Analysis, Computer Methods in Applied Mechanics and Engineering **192**, 3619–3655 (2003).
2. W.S. Wang, Z.H. Fong, A Dual Face-Hobbing Method for the Cycloidal Crowning of Spur Gears, Mechanism and Machine Theory **43,** 1416–1430 (2008).
3. K.H. Chen, C.J. Chen, Z.H. Fong, *Computer Simulation for the Cutting Process of a CNC Hobbing Machine*, Chinese Society of Mechanism and Machine Theory Conference, Chia-Yi, Taiwan, 154-161 (2009).
4. D. Witte, Finish Hobbing Crowned Helical Gears without Twist, Gear Solutions Magazine, 12-13 (2006).
5. J. Lange, *How Are You Dealing with the Bias Error in Your Helical Gears*, AGMA Fall Technical Meeting (2009).
6. O. Winkel, New Developments in Gear Hobbing, Gear Technology, 47-55 (2010).
7. R.H. Hsu, Z.H. Fong, *Novel Variable-tooth-thickness Hob for Longitudinal Crowning in the Gear-hobbing Process*, Thirteenth World Congress in Mechanism and Machine Science, Gearing and transmissions, Guanajuato, Mexico (2011).
8. V.T. Tran, R.H. Hsu, C.B. Tsay, *A Novel Finish Hobbing Methodology for Longitudinal Crowning of a Helical Gear with Twist-Free Tooth Flanks by Using Dual-Lead Hob Cutters*, ASME 2014 International Mechanical Engineering Congress & Exposition, Montreal, Quebec, Canada. Paper No. IMECE2014-36149 1-9, (2014).
9. V.T. Tran, R.H. Hsu, and C.B. Tsay, Study on the Anti-Twist Helical Gear Tooth Flank With Longitudinal Tooth Crowning, ASME Journal of Mechanical Design **136** (2014).

10. F.L. Litvin, A. Fuentes, *Gear Geometry and Applied Theory*, second ed., Cambridge University Press (2004).

MATEC Web of Conferences **52**, 03002 (2016)
DOI: 10.1051/matecconf/20165203002

In-Vitro Design Protocol: Artificial Situation Strategy Uses to Comprehend Designers' Thought

Rusmadiah Anwar[1,3,a], ShahrimanZainal Abidin[1,2] and Oskar Hasdinor Hassan[3]

[1]*Formgiving Design Research Group, Faculty of Art & Design, UiTM, 40450 Shah Alam, Malaysia*
[2]*Research Innovation Business Unit, UniversitiTeknologi MARA, 40450 Shah Alam, Malaysia*
[3]*Department of Industrial Ceramic, Faculty of Art & Design, UiTM, 40450 Shah Alam, Malaysia*

Abstract. The challenge of current trends in design research and point out some of their activities, such as the gap between aesthetic and technical need, and the chasm between ambiguous and quantified in design practice. So, it is hard to surprise the product designers fail to combine a systematical methodology in analyzing design activity. This paper will assess the comprehensible and abilities of In-Vitro Design Protocol (IVDP) in mapping and analyzing design activities. This paper thus attempts to provide a systematic design investigation to researcher and discuss a few key demands on such foundation, include the capture and monitoring used for observational studies in artificial environment strategy.

1 Introduction

Recording design activity among creative designer through video becoming more popular, and define as primary field materials that treat as collecting inquiry data [1] [2]. Even this empirical research has widely practiced among psychologist include engineering designer, it is, however, difficult to find related research involve a kind of research in art and design practices. The interesting about this evidence-based data is, it allowed us to investigate the design thought among chosen subjects empirically. It also brings a mystery outcome during the experiment, which unexpected before. In Valkenburg's recorded videotape and written transcript has shocked by the amount of incomprehension design content among the subjects [3]. If a kind of systematic design methodology such engineering can bring unexpected results, we believed, more intuition design thought from creative designer could be captured and recorded through empirical research. Giving great design behavior and the knowledge gap to be described, particularly in the early phase (the design activity) of any industrial design practices. Where in this stages, the form of a product is poorly supported, mostly at the stage of conceptual [4]. To select a valuable conceptual design, the solution searching process should begin with a divergent until convergence step [5-7].

Conventional models by Cross describe a traditional design process attempt to build prescriptive models of the design process [8]. Cross mostly disagrees with the generation of solution concept identified by these existing models and discovered plenty of examples of the excellent solution provided in a wrong direction. Therefore, Cross suggests an essential structure to the design process of analysis-synthesis-evaluation that not followed the conventional design practice. It was emphasizing

[a] Corresponding author : rusma935@salam.uitm.edu.my

on the performance specifications logical derived from the design problem, generating several alternative design concepts by building-up the best sub-solution and making a rational choice of the best alternative design. This process has same interpretation with "goal-oriented problem-solving activity" [9]. It defines as 'the goal' plays a huge influence any goal-oriented activities. Therefore, the success of any design development depends on the particular establishment at the beginning of the 'real' project goals. Consequently, completeness of design depends on the ability to introduce the legitimate objectives. In a similar point of view, Pugh define that industry was aware of 'total design'; "the systematic activity necessary for the identification of the market/user need, to the selling of the successful product to satisfy that need an activity that encompasses product, process, people and organization" [10].

2 A review of design research related to design activity

Design research is systematic inquiry whose goal is knowledge of, or in, the embodiment of configuration, composition, structure, purpose, value, and meaning in man-made things and systems [11]. However, in current art and design practice, there has been increasing debate about research as a discipline in art and design [12]. Neither artist nor designer is regularly undertaking formal research based on existing research method from any Physical and Social Sciences. Conversely, it does not convey to any of process correctly. Do the art activity (painting, sculpture, printmaking, etc.) involves 'design' or, any design development include 'art' (color, line proportion, scale, etc.) in the practice? It is the nature of art and design and becoming one of the topics for such debates. However, industrial design is not art, craft or engineering; it has it own purpose, values, measures and procedure [13]. Looking at design research from the design methodology and design science perspectives in a sense for such topic, it is principally (five elements of design research) tries to answer the obligation of design to the humanities [14]. It shows that design research only focuses in comprehending the methodology, thus, metaphor structure is neglected and overlooked. The reasons Bayazit describe the elements in design research as his disagreement with the art, craft and design people call what they do for art and design "research." Practicing activities when creating a work of art or craft cannot be considered research [14]. However, somehow rather, he agrees that the observable phenomena are the research gap for the external observer to do research into how an artist is working with their creative works.

In our point of view, most of the design activity traditionally is sketch neither line nor form, the construction can totally consider as "research," especially when the activity involves design process. Design-inspire approaches are crucial in the form development process [15]. Designers generate metaphors by captivating attributes from one or more entity before transforming it to product [16]. These inspirations could come from new ideas from designer's previous knowledge and experiences [17]. Designers use metaphor as a strategic approach to design activity [18] in which, metaphors is one of the mechanisms that permits the well-known mental imagery of subjective experiences and judgments based on general experiences that could be neutrally linked [19]. The metaphorical element is important, as it frequently serves as ways in stimulating designer's creativity in the design process. As an artifact that reveals something insightful, conceptual metaphors are used in relating the connotation of separated entities to be conveyed and comprehended, thus allowing creativity during form development of a product [16]. The metaphors also contribute in assisting in different thinking by bridging different domains in design problem solving [20-21].

To make possible for an external observer able to do research into how the artist or "creative" [22-23] designer's thinks, we use an image to create the link (especially on the meaning) between ideas. The interpretations of design research as defined [14],inclusive with a visual metaphor we have suggested are summarized in Figure 1. Tea brewing machine choose as subject to link with Cross's three main categories that involve design research with people, process, and product [24]. The coherent examples of tea brewing machine have employed successively in examining the two structure variation methods [25]. This variation of relative arrangement defines as 'quantified structure' offer a final choice of criteria looked-for by considering the most promising structure [26].

Based on Figure 1, we can realize the topic of design research mostly related to design activity as defined by Cross. Metaphorically, the design praxeology can be studied into the topic of A, B and E; Design phenomenology study suit with the issue of C and D. Even though topic B is the only design research can fall into epistemology design categories, the same argument (B) has fallen into design praxiology categories. It confirms that discussion in the design process (design praxiology) by the designer (design epistemology) required in-depth investigation. Subconsciously answered the practicing activities when performing creative work can be considered as research. The structure of empirical study based on creative designers will explain in the next chapter.

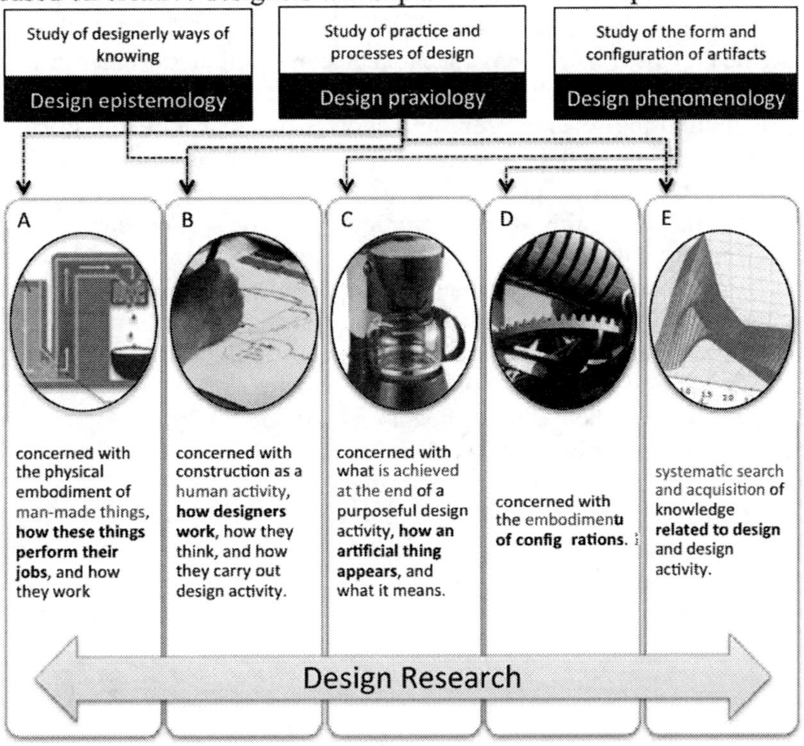

Figure 1. Designer's strategic metaphor design language linked to design research.

3 Organizing in-vitro design protocol as investigation format strategy

As per discussed in Section 2, design grows as a discipline with its research base and Cross believe, there will be a growth in the number of emerging designer-researchers. To devise an appropriate standard or addition to a methodology, it is necessary to identify the 'designerly' ways of knowing, thinking and acting [24]. However, there is no coherent theory exist in current literature about characteristics of metaphorical form in product design embedded based on the representation of mind by designers' from the perspective of Islamic influences. Thus, it become an advantages of this study: (i) to uncover how designers interpret metaphorical form based on Islamic influences, (ii) to understand the influence of incremental, radical design that changes respective to preceding designs, and (iii) to confirm the similarities of characteristics of metaphorical form representation in the context of syntactic that can be generalized as a brand and identity for Islamic product design. We suggest in-vitro design protocol (IVDP) to investigate the existence of metaphorical understanding instead of structure principal during design development. A video observation based on protocol design analysis of designer sketching activities by practitioners conducted and strengthened by a semi-structured interview with an expert on the evaluation of selected product architecture in the analysis of representation concerning metaphorical form syntactic in design. The following steps will describe the fragmentation of activity as the primary structure for this IVDP.

STEP 1: Understanding of design taxonomy (the task).Designers may sketch only to a partial degree ("completeness") in metaphorical form process. How, then, are elements metaphorical form by designers on completeness? [27]. This question close to the design models summarize by Blessing through the typology of design models as stage-based and activity-based [28], which realized the models that concerned with morphological and problem-solving dimension. Furthermore, the model prescribing well-structure activities within stagesand signify to convergence on a design solution where concrete activities within a stage used progressively [28]. The models presented will be as a benchmark for the comprehensively understand the design thought during the experiment.

STEP 2: Time framing (the design abstraction limitation). Even though the typology of design model able to describe the potential convergence to a design solution, designer (after this call 'subject') involves in this IVDP been control by time duration. The task given is to identify as much as solution ideas as possible. Including the selecting multiple choices of ideas that should be considered during design activity [29]. As design activity classified and not think as research due to the demanding reasonable research outcomes while performing an artwork or design work, we decide to bring a "short-term memory" (STM) in this design protocol. The reason to carry STM in IVDP is to investigate how possible the data inquiry during design activity is constructed in a short period. It will be a part of completing goal-targeted action among designers. Tjalve rationalizes the search for examination of solution, evaluation and choice of shows in the theory of search for and examination of solution, evaluation and choice as the characteristic sequence in product synthesis [26].

STEP 3: Conducting design protocol analysis (the experiment). The final formats for IVDP as investigation strategy following the design protocol methodology proposed by Anwar [30]. The issues on matters of the complexity of design research studies, including the difficulty of disentangled the relationship connection, it is important to adopt the same protocol to bring a wider pattern of observation strategy (see Figure 2a). It is an important method to empirically creative designers thought; four cameras (views), influence panel (optional four), clock (time recording), work-space (designer's working area) and sketch book (design activity). The divergent convergentof design brief strategies as visualised in Figure 2ballow us retaining the experiment design format.

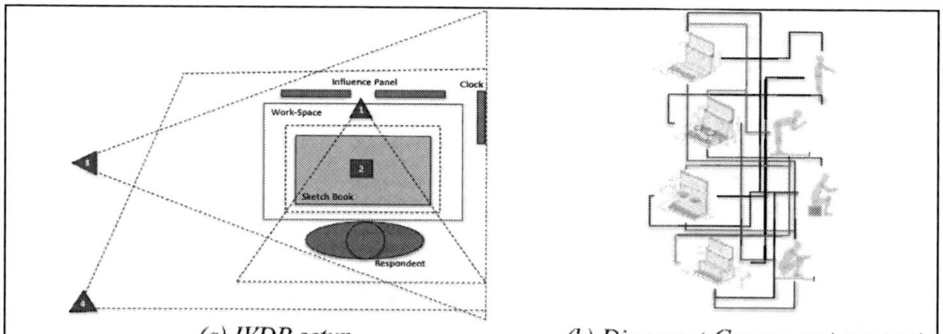

(a) IVDP setup *(b) Divergent-Convergent concept*

Figure 2. Controlled experimental and design protocolreplicated from Anwar et al. [29].

4 The procedure of function means analysis as design strategy

The Ablution Function Mean Analysis [7] set as a guideline on how designers practically apply while constructing form design. The design brief emphasized to analyze and evaluate the formgiving design in a variety of solution. This format is used to be as a document to integrate the design task of constructing Muslim's ablution product concept. Here, subject (designer) must allow to select all four ablution principles, perform on two combinations of four models mentioned with optional of two human posture (see Fig. 2b). These selections decide based on the primary ablution requirement [31]. In this guideline, the selection process known as Concept Integration Approach. This strategy aims to visualize explicitly design approaches through morphology analysis process concerning manual sketching.

We also consider the limitation of time for the subject to develop a concept in short time procedure. As Abidin discover the designers typically use to choose what element to morph rather (identify as "selectivity") than transforming consistently [32], we set four panels (see Table 1) as an influence on the subject to 'spark' their ideas. It has divided into four groups; (Panel 1) Icon, Sign, and Symbol; (Panel 2) Object and artifact; (Panel 3) Building and Architecture; and (Panel 4) Culture and Heritage. The selected image for all panels is based on "public domain" with all images signify to Islamic characteristic. The purpose of this influence panel is to identify rather this picture will sketch in explicit or implicit ways. To investigate what expression portrays, the subject is asked to reflect on the question we provide. The following section determines how designer think in such a way introducing a basic structure for a product.

Table 1. The influence panel

Four Types of image randomly selected as Influence Panel			
Panel 1	Panel 2	Panel 3	Panel 4
	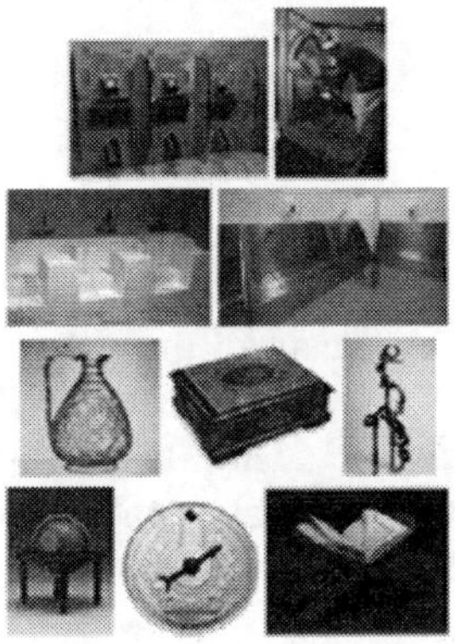		
Icon, Sign & Symbol	*Object &artifact*	*Building & Architecture*	*Culture & Heritage*

5 Analysing designer's thought throughout IVDP experiment

Experiencing the IVDP as illustrated in Figure 3, the ambiguous characteristics of metaphorical form through designers sketching processes of Islamic product design has lead to a natural variety in output. Whereas Abidin refer this phenomenon as "consistency." Thus, how do designers assess metaphorical form through their sketching assignments has discovered [18]. Throughout thirty minutes of design activity (short-term memory), this empirical study stipulates in-depth qualitative data reflected all artificial situations that have been arranged.

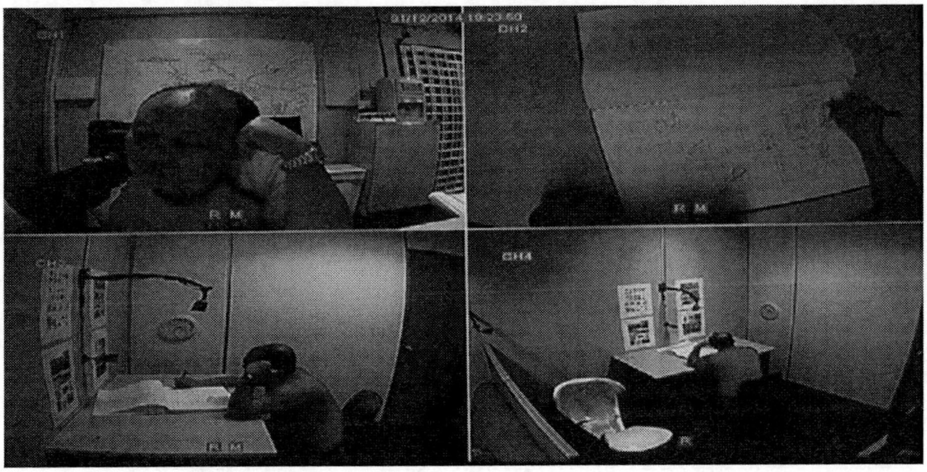

Figure 3. Expert designer as 'subject' during IVDP experiment.

At the beginning of design activity, the subject often uses morphological analysis as similar with Funtion-Means Analysis Chart (See Table 2). A morphological chart was originally developed as a design method, overviews of function considered and representation produce especially on the product design perspectives. The morphological analysis has allowed subject to analyze the development procedure of ablution design concept. From here, the subject used to "trace out" potential whole system concept solution. Table II shows how the subject accomplishes the analysis by referring the Function Means Analysis for Ablution.

Table 2. The Function Means Analysis chart applied by the designer during the creation of ablution design [7]

Function - Means Analysis Chart				
Function	Means			
Ablution Principle	Face	Both hand	Forehead	Both Feet
Body Posture While Ablution	Stand	Bend	Sit	Squat
Faucet design	Lever-type	Single-hole	Center-set	Wall-mount
Water fall method	Flush (Pipe)	Dipping (Pond)	Scoop (Bailer)	Bailer & Dip (Pond)
Drain System	Open	Covered	Pipe	Trap & pipe

IVDP experiment allows us to identify that designer able to derived design and divide three different levels of abstraction, which is abstract, semi-concrete and concrete. On the first level of design development (Table 3a), the subject seems to illustrate the form construction by tracing out non-priority means for ablution. Remaining means offer an initial and reasonable idea in realizing the concept. Structure design brief by stating the direction of design task in an apparent divergent phase, acknowledge by subject to present an excellent level of convergent. From the early (abstract level) process of design concept, the subject has determined the visou-spatial characteristics [33] of the ablution product. The form structure and configuration completely close to 'higher order form elements' [34]. As refers to the abstract level, the subject has "assemble" all form element into one structure, with the inclusive of Islamic (arabesque) characteristic. Undergoes to the next level (semi-concrete) of a design process, the subject chose to implement a quantified structure strategy (See Table 3b) to determine the best consideration and most promising design structure for ablution practice.It comes to the end of design activity (concrete level) as subject confidently configures the form element with the relation of ablution requirements. In Table 3c, though the product development, the subject has determined a gestalt study [35]. The ablution design structure organized, arranged and correlated well with the requirement of ergonomic, technical and practical of ablution performance requirements.

Table 3. The abstraction level of design experiment during IVDP

The Abstraction Levels		
(a) Abstract	(b) Semi-concrete	(c) Concrete
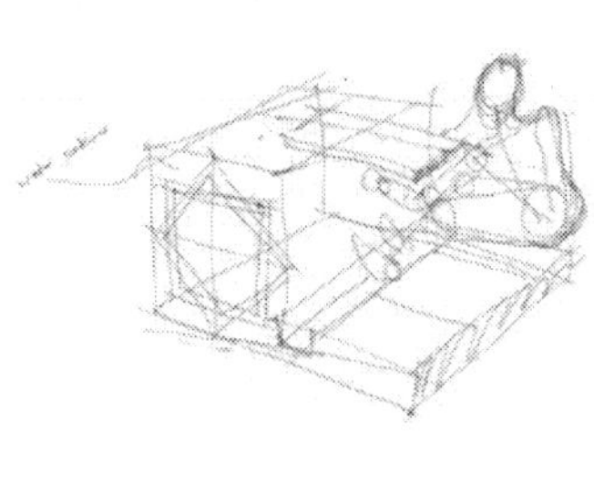	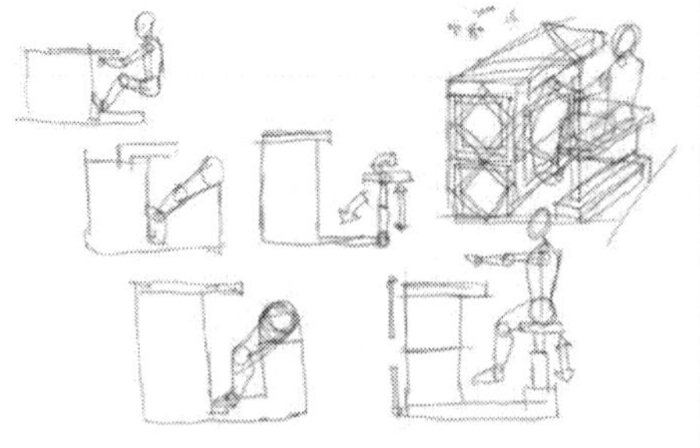	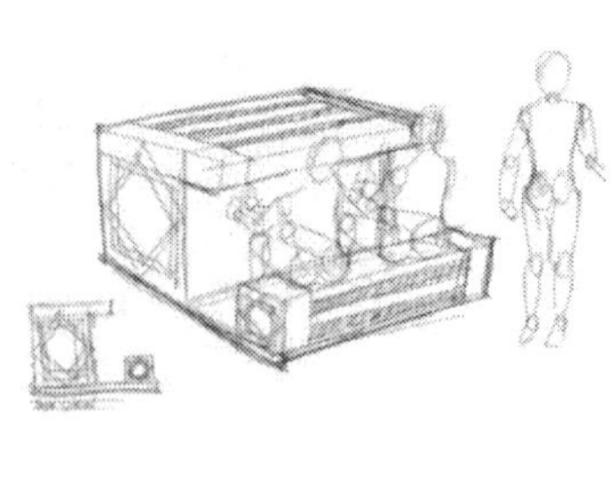
The assembly of form elements.	*Quantified structure of form elements.*	*The configuration of form elements.*

6 Discussions

In this research, the design activity performed by creative designer able to give a contribution to the body of knowledge, especially in exploring how creating a craftwork (design) canrelateto the field of design research. It has proven in anyabstraction (abstract/semi-concrete/concrete)level of the design process has metaphorically give a significant meaning as research investigations. As the IVDP experiment is artificially set as design situation to provoke designer in constructing a form, the subject spectacularly drives the way of design think and approach into 'Formgiving Design Principle Solution' as a way for problem-solving. Transforming the whole design process into a concrete idea development stages. As 'Product Architecture' is a method whereby; the functional elements of a product will be assigned to its physical building blocks. Moreover, it can be implemented as early as idea development takes place, through sketches, function diagrams, and early sketch models [36]. So we refer to the experiment (Divergent Convergent Process) through IVDP as the first stage of next investigation strategy (product architecture) take place. IVDP has offering design investigation format, and it recalls on how practicing activities by a creative designer can considered as systematic research and given a cues to support another research phase.

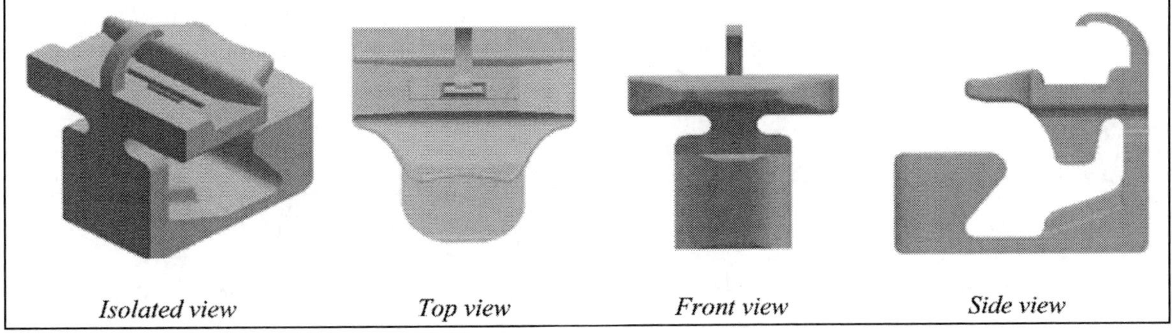

Figure4. The conceptual and embodiment form element of ablution throughIVDP experiment.

The experiment mean to focus on Divergent Convergent Process strategically to develop idea inspiring complete form of ablution concept (Figure 4). Throughout subject's 'function means analysis', we construct completely prefabricated idea (form) to make possible the concrete idea transformation through abstraction level (see Table 3) as a sample of a structure to attempt another phase of a design process. Especially when it involve the concept development of Product Architecture, which determine by product change, product variety, component standardization, product performance and product development management [37]. Through our analysis, the design structure(syntactic) formed during the abstraction level (Table 3a, 3b, and 3c) has summarized, it is important to structure explicitly the ablution to occupied the need to perform the principles (face, both hand, forehead and both feet). However, the designer in such a way has practical means on the forming surface of the form (design structure) in representing the influence of their intuition mind strategy. The only visible element used as an explicit design is the arabesque design. Three level of abstraction implies this element to represent the product was initiated by Islamic influence.

We hypotheses that the activities of interaction with influence panel make the designer matured with a stronger embodied understanding of the form structure. This IVDP experiment also shows that designer interacts the mental image with eye contact and experience in designers' apparently intuitive formgiving process. The designer has played with particular 'configuration system' in an intuitive stage to understand and visualize form related to the image in the mind. Function Mean Analysis here uses to identify on, how well the designer knows when is the right time to change from one domain to the other. The transformation of abstraction level(through designer's think and working drawing) clearly through conceptual and embodiment phases represents a product-based system, and allowing us to decompose the analysis system due to the similarity in term of design outcomes.

7 Conclusions

In conclusion, we strongly disagree if creative work cannot be considered as research and only taking apart of observable phenomena. Cross, in his taxonomy, agree that three main categories should be a part of design research which based on people, process, and products. In our first finding, the design practices or topics mostly have a strong connection to design research. Three out of five research topic mentioned used by the designer as key metaphor design categories. IVDP as artificial situation strategy with well structure design brief has stated clear direction of design task and how designer's think while creating a creative work. The designer where in this study we call as 'subject' has proved that design development constructed during 'abstract' 'semi-concrete' and 'concrete' level, often close to 'goal-oriented problem-solving activity' as defined by Archer. It shows that even creative designer can offer a systematic investigation, especially through IVDP. Novel facts, using a scientific method such IVDP may provide a new method of design research especially in the field of art and design.

Acknowledgement

We would like to acknowledge the generous participation of the interaction designers in the research. This study was conducted in Formgiving Design Research Lab, UiTM. Fully appreciation to Malaysia Ministry of Higher Education for the financial support under RAGS and UiTM for the REI grant.

References

1. S. Z. Abidin, D. Christoforidou, and A. Liem. Thinking And Re-Thinking Verbal Protocol Analysis In Design Research. Stanford: International Conference On Engineering Design, pp.1-12(Aug. 2009).
2. R. Anwar, S. Z. Abidin, and O. H. Hassan. A Framework of Empirical Study Through Design Practice for Industrial Ceramic Sanitary Ware Design. In ICADER 2014 by O. H. Hassan, S. Z. Abidin, R. Legino, R. Anwar and M. F. Kamaruzaman (eds). Singapore: Springer (2015).
3. R. C. Valkenburg, Shared understanding as a condition for team design." Automation in Construction, vol. 7 (2), pp. 111-121 (1998).
4. B. Sener, and P. Wormald. User evaluation of HCI concepts for defining product form. Design Studies, vol. 29 (1), pp. 12-29(2007).
5. G. Pahl, and W Beitz. Engineering design. London: The Design Council and Springer Verlag(1984).
6. S. Pugh, D. Clausing, and R. Andrade. Creating innovative products using total design. New York: Addison-Wesley Publishing (1996).
7. R. Anwar, S. Z. Abidin, and O. H. Hassan. Function Means Analysis For Ablution Concept Solution. The Turkish Online Journal of Educational Technology, pp. 224-231(2015).
8. N. Cross. Engineering Design Methods: Strategies for Product Design. Chichester: John Wiley & Sons, Ltd (2000).
9. L. B. Archer. Systematic Method for Designers. In Developments in Design Methodology, by Nigel Cross (eds). Chichester: John Wiley & Sons(1984).
10. S. Pugh. Total design: integrated methods for successful product engineering. Boston: Addison-Wesley(1991).
11. L. B. Archer. A View of the Nature of the Design Research. In Design: Science: Method, by R. Jacques and J. A. Powell, Guilford, Surrey: IPC Business Press Ltd., pp. 30-47(1981).
12. C. Frayling. Research in art and design. vol. 1 (1), London: Royal College of Art Research Papers(1993).
13. C. L. Owen. Design research, building the knowledge base. Design Processes Newsletter 5 no. 6, pp. 1-6(1994).
14. N. Bayazit. Investigating Design: A Review of Forty Years of Design Research.Design Issues 20 (1), pp. 16-29(2004).

15. S. Z. Abidin, J. Sigurjonsson, A. Liem, and M. Keitsch. On the Role of Formgiving in Design. International Conference On Engineering & Product Design Education. Barcelona, pp. 365-370(Sept. 2008).
16. N. Cila, P. Hekkert, and V. Visch. Source Selection in Product Metaphor Generation: The Effects of Salience and Relatedness. International Journal of Design, vol. 1 (2), pp. 15-28(2014).
17. R. L. Marsh, T. B. Ward, and J. D. Land. The inadvertent use of prior knowledge in a generative cognitive task. Memory & Cognition. vol 27 (1), pp. 94-105(1999).
18. S. Z. Abidin, H. V. Bjelland, and T. A. Øritsland. The embodied mind in relation to thinking about form development. Proceedings of NordDesign 2008 Conference. Tallin: pp. 265-274, Aug. 2008.
19. G. Lakoff and M. Johnson. Philosophy in the Flesh The Embodied Mind and Its Challenge to Western Thought. New York: Basic Book, 1999.
20. K. Dorst. The Core of 'Design Thinking' and Its Application. Design Studies. vol. 32 (6), pp. 521-532(2011).
21. H. P. Casakin. Factors of metaphors in design problem-solving: Implications for design creativity. International Journal of Design. vol. 1 (2), pp. 21-33(2007).
22. G. Pahl, W. Beitz, J. Feldhusen, and K. Grote. Engineering Design: A Systematic Approach. 3rd. Edited by K. Wallace and L. T. M. Blessing. London: Springer-Verlag(2007).
23. R. Anwar, S. Z. Abidin, and O. H. Hassan. A Pattern in Formgiving Design: Giving Priority to a Principle Solution in Industrial Design Situation. In Industrial Engineering, Management Science and Applications 2015, by M. Gen, K. J. Kim, X. Huang and Y. Hiroshi, Berlin Heidelberg: Springer, pp. 331-340(2015).
24. N. Cross. Designerly Ways of Knowing. London: Springer-Verlag(2006).
25. V. Hubka, M. M.Andreasen, and W. E. Eder. Practical Studies in Systematic Design. Edited by Peter Hills. London: Butterworths(1988).
26. E. Tjalve. Systematic Design of Industrial Product. Lyngby: Institute for Product Development(2003).
27. S. Z. Abidin. Practice-based design thinking for form development and detailing. PhD Thesis. Trondheim: Norwegian University of Science and Technology, 2012.
28. L. T. M. Blessing. A process-based approach to computer-supported engineering design. PhD thesis. Twente: University of Twente, 1994.
29. R. Anwar, S. Z. Abidin, and O. H. Hassan. Understanding Methodological Solution In Design Situation Of Novice Designer. 18th International Conference on Interactive Collaborative Learning, Florence, IEEE. pp. 593-597 (Sept. 2015).
30. R. Anwar, S. Z. Abidin, and O. H. Hassan. The Turkish Online Journal of Educational Technology, pp. 13-21(2015).
31. N. H. Johari, R. Anwar, and O. H. Hassan. "Design framework of ceramic ablution Tub." IEEE Symposium on Business, Engineering and Industrial Applications, Bandung, IEEE, pp. 608-610(2012).
32. S. Z. Abidin, A. Warell, and A. Liem. Understanding Styling Activity Of Automotive Designers: A Study Of Manual Interpolative Morphing Through Freehand Sketching. Proceedings of the 18th International Conference on Engineering Design (ICED11). Copenhagen: Design Society, pp. 357-366(Aug. 2011).
33. W. Muller. Order and Meaning in Design. Utrecth: Lemma Publisher(2001).
34. A. Warell. Design Syntactics: A Functional Approach to Visual Product Form. Goteborg: Chalmers University of Technology(2001).
35. R. Mono. Design for Product Understanding: The Aesthetics of Design from a Semiotic Approach. Stockholm: Liber AB(1997).
36. A. Liem. Managing The Industrial Design Process: A Guide for Studio Practice. Singapore: Prentice Hall(2004).
37. K.T. Ulrich and S.D. Eppinger. Product Design and Development, 5th, New Delhi: TATA McGraw-Hill Edition(2005).

MATEC Web of Conferences **52**, 03003 (2016)
DOI: 10.1051/matecconf/20165203003

The Application of Cloud Innovation in Industrial Design Education

JIANG Li-jun[1], Song Bo[2], Su Jing[3], Li Zhe-lin[4,a]

[1] *School of design, South China University of Technology, Guangzhou 510006,China*
[2] *School of design, South China University of Technology, Guangzhou 510006,China*
[3] *Foshan 91Ant network technology co., LTD, Foshan 528000, China*
[4] *School of design, South China University of Technology, Guangzhou 510006,China*

Abstract. According to the domestic status of students' study and practice of industrial design, the application of cloud innovation in industrial design education was proposed to realize online design reviews, dissemination and sharing, virtual display and demand-oriented collaborative design. The basic framework and function of cloud innovation system were analyzed to solve three problems: design relearning, lack of creative communication, separation between education and enterprise production of industrial design.

1 Problems of Learning and Practicing in Industrial Design Education

The necessity of industrial design innovation is not only passion, intuition and imagination, but also the strong rationality and the right approach. From the practical needs of industry, oriented from the problems, industrial students need to balance creativity with other constraints by design evaluation and design iterations in various stages. Restricted by the domestic objective factors, students of industrial design have many problems in learning and practicing. It is the key to the development of industrial design that how to develop students' creativity effectively combined with the concept of Internet+.

With the popularity of Web2.0 technology, cloud innovation comes to represent the most viable innovation model, which concentrates wisdom of public innovation. To solve the problems of learning and practicing in industrial design education, the needs of the students' professional development were analyzed and the cloud innovation platform was set up and studied.

On the basis of analysis research, three problems were summed up for industrial design students when learning and practicing.

1.1 Relearning problems with new design task

As industrial design is interdisciplinary, designers have to design the product in the form, style and emotion with knowledge about materials, technology, beauty, and culture, to realize product innovation with high technology, artistic value and economic worth(as shown in figure 1). It means that the more interdisciplinary knowledge structure, more sophisticated technical support and better innovation theory are needed. Industrial design students encounter relearning problems with new

[a] Corresponding author : gzwoods@qq.com

design task. To save learning cost, students need the guidance from experienced people, or team of people with variety knowledge and technical background during the design process.

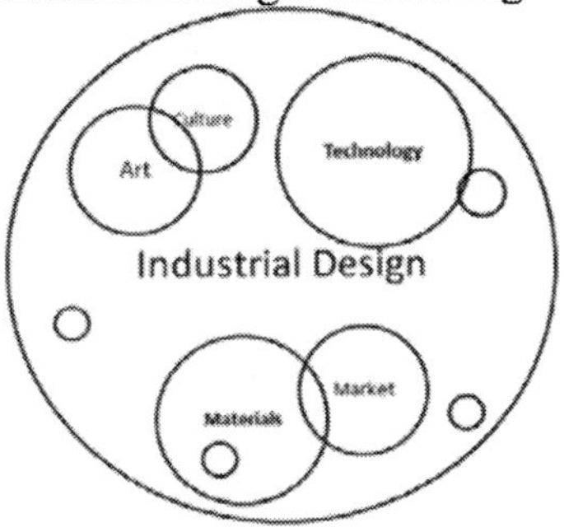

Figure 1. The Relationship Between Industrial Design And Relevant Subjects

1.2 Lack of innovative communication

The major of industrial design is set up by many colleges and universities in China. A lot of design works, some of which are quite creative, are sealed in graduation documentation; At the same time, the shortage of design talents for industrial enterprises and design companies need to be solved and a large number of design students cannot obtain employment. Due to the limitation of objective and subjective elements, communication between enterprises and teachers, students from different colleges is not sufficient. The existing closed innovation model is not suitable for most designers.

1.3 Education separated from practice

In universities at home, the design theory domestic students learning is relatively old sometimes, and the design practice they make differs from the actual production conditions. Due to the geography and other objective conditions, many students don't have an opportunity to link with companies and only do virtual design exercises at school. As the students can not participate in the the real design process, their design works in school are hard to turn into real products.

2 Research of the Existing Design Innovation Platform

For now there are many online design innovation platforms for business and education in domestic and overseas. From the perspective of industrial design education, they all have advantages and disadvantages (examples are shown in table 1).

Table 1. Design Innovation Platform Research

Platform name	Platform type	Basic mode and functions	Target audience	Advantages in design education	Disadvantages in design education
Billwang (www.billwang.net)	Industrial design communi-ty	To publish and discuss design work,get information of design activities and recruitment	Industrial designersin dustrial design students	Professional and targeted	As most students' design are in a primary level, their design works will not gain enough attention and advice
ZBJ Network (www.zbj.com)	Witkey Service Platform	Employers post requirements, providers bid and work on jobs posted by employers. Three transaction-al models: reward, bidding and trading patterns	Enterprises, social organizatio n-s and individuals; Witkey, design insti t-ution	Through design practice, students can have better knowled-g e of the market requirements	Design work is only judged by the employers alone, without design professionals and normal users' opinions, there are restrictions on the improvement of students' design

					capabilities.
Kickstarte-r (www.kickstarter.com)	Crowdfu-nding platform	To help bring creative projects to life before the deadline by funding money from public	Creative people who need money for projects	There are complete project processes and high innovative design solutions; Interactive communica tion between users	The main objective is to promote the realization of projects, not to raise the design level
Behance (www.behance.net)	Design communi -ty	To showcase and discover creative work, also to look for designers	Designers, design students and other design enthusiasts	Students can learn from the complete and high-quality designs on it	As there are many outstanding designers, students' design works are hard to gain enough attention
Coursera (www.coursera.org)	Massive open online courses	To take free online classes from top universities and educational organizations	Learners, including university students	High-quality design courses, and assessments mentor and facilitat-e learning process	Less teacher-student interaction and less opportunity to practice, which is quite necessary for industry design study

The analysis above shows that the existing design learning platforms have their own advantages, but not perfect for industrial design education. The existing imperfection shows that the theory and method of cloud innovation may make up for these above disadvantages.

3 Concept of Cloud Innovation And Related Studies

Cloud Innovation is a wide range(usually worldwide) innovation activity which reflects the swarm intelligence[1] .It connects various of knowledge and talents from worldwide to set up an normalized innovation service mode which combines scattered innovation resource. It is appropriate to be applied to students' online education platform.

Domestic and foreign researchers proposed many effective and innovative ways about learning methods of innovative design, such as expert systems, learning systems, mind mapping, brainstorming, etc. Currently, the innovative design theory, research methods and tools, reflecting the trends from closed to open[2-3].

Overseas, such as researcher Steven Tao in Carnegie Mellon University studied how to apply crowd sourcing methods to traditional education mode. If this is an appropriate way, it will allow many workers to work in a collaborative manner through the Internet, and make a way to generate income for the US economy[4].

In domestic, School of Design in South China University of Technology studied the communication of members in community and sharing of knowledge, which proposed mechanisms model of knowledge sharing in virtual design community from the perspective of teamwork[5]. School of Management in Huazhong University of Science and Technology discussed the propulsion mechanism of KIBS open innovation in three areas: creation, transition and diffusion, knowledge dissemination and absorption capacity[6].

At present, researches of cloud innovation are still in primary stage. The application of cloud innovation hasn't form a systematic theory in industrial design learning and practice, which has a huge exploration space.

4 The application of cloud Innovation in industrial design

To solve the above problems, based on the cloud innovation theory, we need a platform to realize online design reviews, dissemination and share, virtual display and demand-oriented collaborative design, to help students study and practice in industrial design better. Consequently, 91Ant[7] was built to validate the idea.

4.1 Framework and functions of cloud Innovation platform of industrial design

4.1.1 Design work show

Cloud innovation platform cooperates with design schools. Students are be encouraged to upload their works of design courses and exercises. Teachers and students from different regions and different schools can do the reviews and discussions about the design online easily without long distance-travel.

Figure 2. Page of design work show and page of uploading

4.1.2 Design exhibition and competition

Online graduate design exhibitions and design competitions are held to gain new users and design topics enhance user activity. Ordinary users are also invited to share their using experience and ideas of the product(as shown in figure 3), so that designers will understand what customers need. Moreover, students get more better ideas through online brainstorming and other methods. The problem of relearning and lack of communication is solved through the above ways.

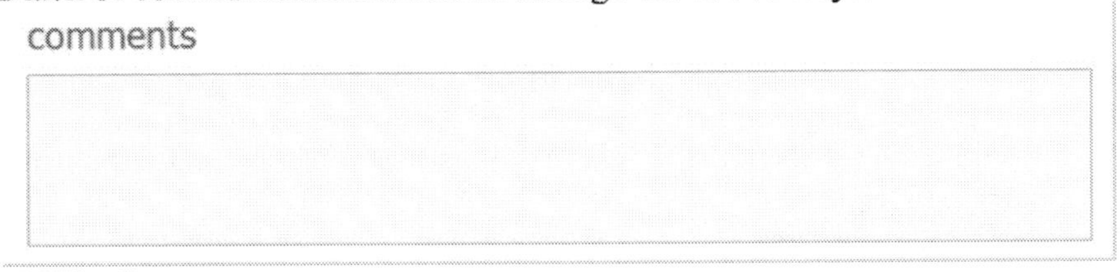

Figure 3. Page Of The Comments

4.1.3 Practice in cooperation with enterprises

Cloud innovation platform cooperates with enterprises. Enterprises call for designs of their industrial products on platform(as shown in figure 4) . Students who accepted the proposition should upload the innovative ideas or preliminary design to the platform. Comments and discussions are encouraged on the platform, which are made by students and teachers of design, experienced industrial designers, and people of other relevant disciplines (such as materials science, processing technology, market, businesses,etc). Students adopt reasonable opinions and improve their design. Companies adopt the good ones and pay for them. Members participate in the discussion will get reward points depending on the contribution for the design. (as shown in figure 5) The problem of separation between education and practice is solved.

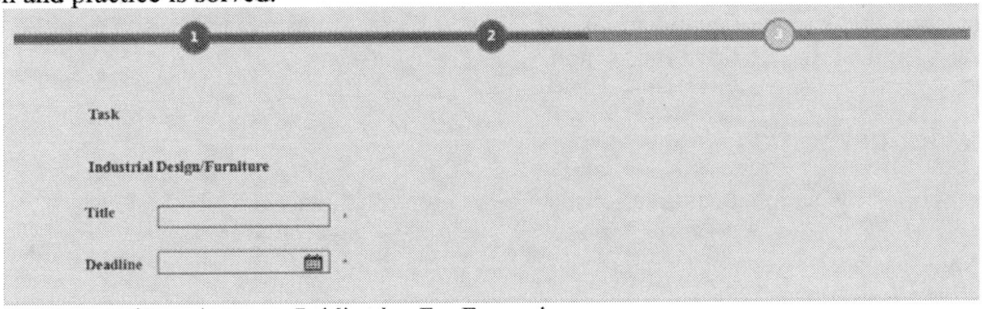

Figure 4. The Page Of Requirements Publication For Enterprises

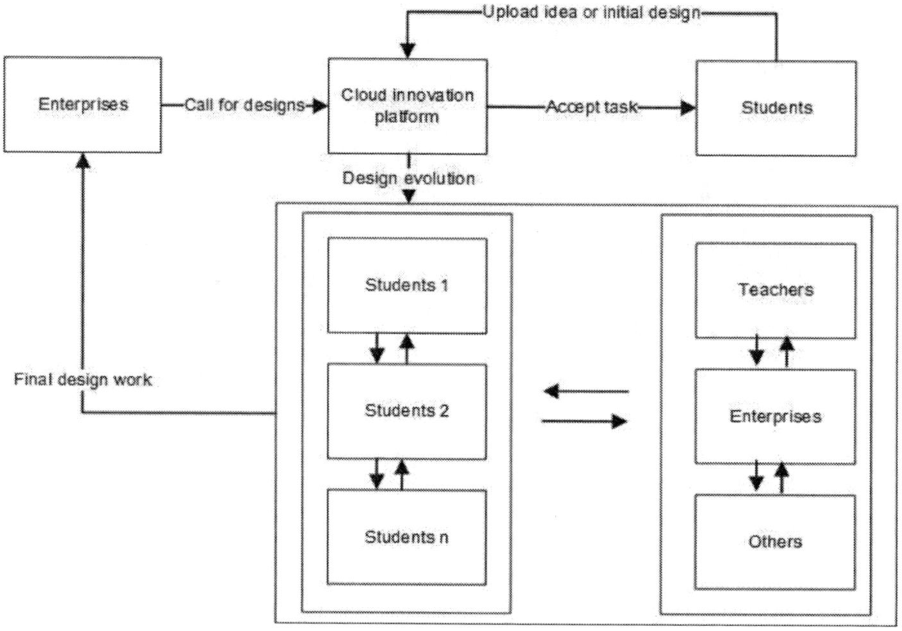

Figure 5. The Process Of Design On Cloud Innovation Platform

4.1.4 Others

In addition, video teaching of college courses, enterprise courses and master's lecture is also provided to supplement the theory knowledge.

In cooperation with the National Copyright Administration, cloud innovation platform protect students and designers' copyright.

Up to June 2015, the scale of China's mobile phone users reached to 594 million. As a consequence, some functional modules will be developed under mobile network, too. The convenience of mobile networks will greatly enhance the breadth and efficiency of dissemination.

4.2 Effect of cloud Innovation platform of industrial design

Since the establishment of 91Ant in December 2015, 732 design works has been uploaded, and design graduation show organized by Guangdong Vocational and Technical College is also welcomed. We got some user survey feedback:

Student A: From 91Ant, I got a better understanding of the needs of market.

Student B: I can visit the graduation show of other universities at home now.

Student C: This is a platform designed for students, so, our design won't be ignored here.

The monumental achievement of 91Ant demonstrated that cloud Innovation platform had some positive effect in the industrial design education for students in some degree. However, because the time of evaluation is relatively short, further studies should be needed to verify the applicability of cloud Innovation.

Acknowledgements

Project supported by National Natural Science Foundation of China(No.51 1 751 83)

5 Summary

Cloud innovation is a key to promote creativity, and the application of cloud innovation in industrial design education is possible to solve the problems in students' study and practice. It is a new exploration and has an important significance to promote Chinese industrial design education. However, there are still many shortconmings and omissions in this paper, and more evidences are needed to prove the theories and ideas.

References

1. Ren liMei, Huang Bin, Cloud Innovation-innovation model in the 21st Century[M]. Beijing, Chinese Communist Central Party School Press
2. TIAN Qi-hua, XIAO Ren-bin, ZHONG Yi-fang, DU Yi-xian, YANG Hong-mei, Integration of axiomatic design and theory of inventive problem solving for conceptual design [J]. Journal of Chongqing University (English Edition) [ISSN 1671-8224] (2009), 8(3):209-216
3. Jacob Goldenberg, David Mazuisky, Creativity in Product Innovation[M]. Cambridge University Press, (2003)
4. Magy Seif El-Nasr, Jacquelyn Ford Morie, Anders Drachen, A Scientific Look at the Design of Aesthetically and Emotionally Engaging Interactive Entertainment Experiences[OL], http://www.sfu.ca/~magy/conference/Emotions-chapter.pdf
5. Yang Qiang, Research of Organization Construction for Open Cloud Innovation Oriented Virtual Design Community[D]. South China University of Technology master's thesis, Guangzhou, (2014).6-10
6. He Xiaoyu, Cao Yong, Research on promoting mechanism of open innovation in knowledge intensive business service [J], Science and Science Management, (2010), (1): 59-64
7. Www.91ant.com

MATEC Web of Conferences 52, 03004 (2016)
DOI: 10.1051/matecconf/20165203004

Modern Landscape Representation of Hakka Culture

Huang Hui[1,a], Gao Ruofei [2]

[1] *Landscape Design Department, Shenzhen Polytechnic, Shenzhen, China*
[2] *Doctor of Chiba University, Japan*

Abstract. Hakka culture is a combination of ancient Chinese culture, aboriginal southern culture and hilly environment, whose influence spreads more than 80 countries and regions in the world. It is more difficult to summarize the regional landscape representation techniques of Hakka culture under the background of modern landscape. The author makes a comparative study of landscape representation techniques of the three typical Hakka landscapes, putting forward modern landscape ideas of Hakka culture in order to provide reference for construction of other regional landscapes.

1 Introduction

Hakka culture is not only diverse but also unique [1]. Hakka people is a special department that was ultimately formed after several ethnic migrations, which is one of the Han nationality branches with many characteristics of southern ethnics at the settlement surroundings, retaining the relative integrity of the central plains culture. Meanwhile, it is a diverse and unique cultural system that was ultimately formed after the interaction and permeability with the local culture of Baiyue and Yaoshe.

Hakka traditional architecture, especially Hakka Round House, is one of the important carriers and symbols of Hakka culture, which is known as "the world's unique mythological mountain building model [2]". The research, resulting from the study of many Hakka traditional architecture [3] [4] [5] [6] [7], is helpful to summarize the basic connotation and typical characteristics of Hakka culture.

In the modern landscape design, how to inherit local culture better so that the design works are of more regionalism, melting into regional environment, and how to maintain a more long-term vitality, are the ongoing topics of the modern landscape design practice [8] [9] [10]. "World Hakka city" in Meizhou City, Guangdong Province continues to build the three large Hakka scenic areas: Yannanfei tea garden, Ketianxia Tourism Park and Ruishan mountain hi-tech eco-agricultural park. While comparatively studying the landscape representation techniques of the three main Hakka scenic areas, this paper aims to summarize the Hakka culture, trying to put forward the basic concept of modern Hakka landscape design so as to provide reference for the regional landscape design.

1 The Hakka culture

1.1 Hakka culture and its typical characteristics

[a] Author : huanghui72@szpt.edu.cn

Hakka culture springs from the Confucian culture at the year 0 to A.D. 1000, which has developed during five migrations to the South at the year A.D. 1000 to A.D.2000. Mr. Ge Jianxiong said: "Since the purpose of migration is to settle down, people have to make a choice in culture. They can accept the culture which they migrate to so that they can melt into it; or they have to adhere to their own culture so as to carry it out among the local people." However, the Hakka people in the history adhere to the Confucian culture during the migrations, while they gradually integrate into cultures from place to place, which is a kind of derivative culture under the influence or the choice of the natural environment and human environment. Subsequently, due to the relocation to Nanyang*, Hakka culture becomes a model of integration of Chinese culture and world culture.

*Nanyang:a general name used towards the end of the Qing Dynasty for the coastal provinces of Jiangsu, Zhejiang, Fujian and Guangdong; an old name for the Malay Archipelago, the Malay Peninsula and Indonesia or for southeast Asia.

1.2 Hakka culture and Hakka traditional resident community

Hakka Round House is an important carrier of Hakka culture. The building community which is harmony with mountains, the magnificent and simple style, bright and plain color, natural ecological environment and introverted inspirational decoration make it unique among the traditional local dwelling houses.

Hakka Round House faces water with hills on the back, whose terrain falls off the land from high to low in turn: Zu Shan Mountains, Feng Shui landscape forest, the settlement architecture, Wo Ping, Moon pond (Feng Shui pond shaped as half-moon), fields and streams. Ecological base of traditional Hakka residential community is composed of the mountain hills, rivers and streams and plant community. The round house patches, isolated from each other by the terrain, are connected by streets and water body. Such special ecological structure melts the natural elements into the living community, which is a model of the organic combination of the living community and natural environment.

2 Modern representative Hakka landscape

2.1 Three modern Hakka landscape cases

Meizhou City in Guangdong Province, known as the "World Hakka city", is the main ancestral home of overseas Chinese Hakka. Yannanfei tea garden, Ketianxia Tourism Park and the Ruishan mountain hi-tech eco-agricultural park are three relatively large Hakka theme parks that have been newly built at Meizhou in recent years.

The Yannanfei tea garden (Figure 1) is one of the national AAAA level scenic spots, which is a set of Hakka tea culture and tourism with an excellent eco-environment. Based on the superior natural eco-resources and tea fields with standardized production, scenic areas show a broad and profound tea culture. The landscape and architecture in scenic areas, such as thick and Green Tea Park, water,

Villa, Round-Dragon Hotel and Round-Dragon Restaurant, work in concert with the traditional Hakka residential community.

Ketianxia tourism park (Figure 1) was built among the natural mountains with original ecology and the three major reservoirs. By digging deeply in Hakka culture, the park is set up ten major cultural projects and the five major scenic areas, changing "culture pulse" into "landscape pulse" throughout the park, which is an epitome of a combination of history and modern, culture and leisure at Meizhou City.

Ruishan mountain hi-tech eco-agricultural park (Figure 1), taking ecology and health as the theme, is a beautiful place with overlapping peaks and metasilicic acid underground water source which is beneficial to the human body. Being harmony with the mountains, three hotel constructions whose configuration takes Hakkas Round-Dragon house as a reference, constitute a shape of Ruyi with three oval buildings (service center, restaurants and office buildings), which symbols good fortune and longevity.

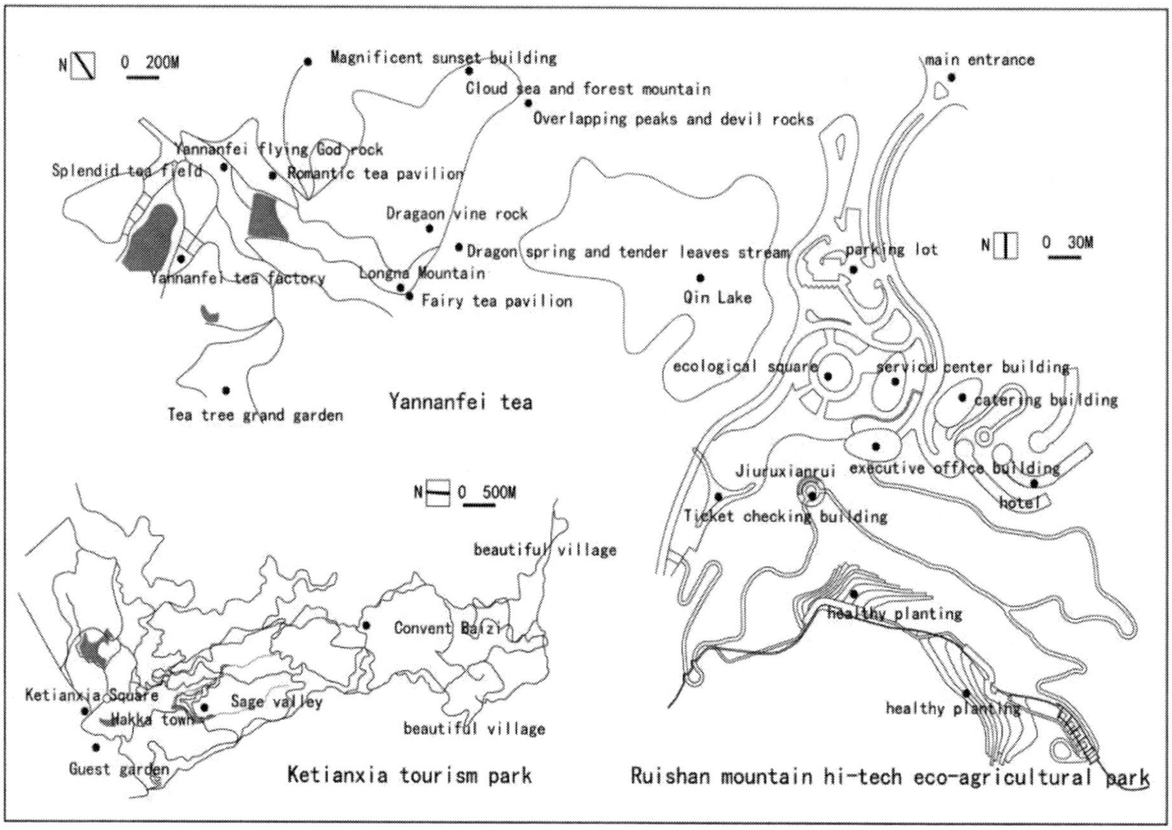

Figure 1 Master plan for Yannanfei tea garden, Ketianxia tourism park and the Ruishan mountain hi-tech eco-agricultural park

2.2 Typical landscape representation techniques

Contemporary landscape designers think that the landscape is a text that can be read and understood, whose major component is the landscape symbol. By selecting landscape symbols, and reorganizing codes and structures, combined with the hints, associations, memories and other techniques, landscape design closely links these obvious or ambiguous symbols and metaphors with history and society, arranging cultural information in it, so as to make people have an assocation that proceeds from one to another so that people can understand things that can not be seen, achieving emotional resonance. From the perspective of the formation of the symbols and structures, there are two following types of refinement of the traditional landscape symbols: general type and restructuring type. General landscape symbol, often originating in people's common cognitives, generally is conventional image and meaning with the characteristics of commonality and universality. The restructuring landscape symbol is a new kind of symbol structures, which is formed by restructuring the materials, patterns or structures of traditional landscape symbols. Being closely connected with traditional culture, the restructuring landscape symbol keeps a tentative relationships with traditional so as to produce the symbol and metaphor for traditional culture.

After field-exploring in three Hakka scenic areas, collecting and reorganizing materials, the author summarizes typical landscape representation techniques as follows (Figure 2):

Figure 2 The basic structure of Feng Shui forest, terraces and streams

Being similar with the patterns of traditional Hakka residential community, the basic structure of the three Hakka scenic areas retain the Feng Shui forest, terraces and streams and other elements. Being different from the proportion of buildings and land in the traditional Hakka settlements, Yannanfei tea garden whose main purpose serves for production focuses on the terrace with structures such as hotels, villas and teaculture buildings as an embellishment. While cultivating the Feng Shui forest, the hotel and the valley streams in the Ruishan Mountain hi-tech eco-agricultural park, the terrace is replaced by the healthy plant in order to strengthen the health function of the park. These general landscape symbols and the landscape structures that are composed of those make it very easy for people to have an association with the traditional Hakka settlements.

1) Round and curved composition and decorations

Drawn from Hakka representative round house, many modern Hakka buildings have the characteristics of resistance of the chill of the valley, enlargement of the utilization area and reduction of sunshine corner, etc., which are more suitable residential patterns for hotels. In addition to hotels, the landscape architecture and decorative patterns are more likely to be circular or arcuate so as to echo the architectural forms. For instance, a large number of circular paving patterns echo to the form of hotel architecture in Yannanfei tea garden. In addition, round Central Square in Ruishan mountain hi-tech eco-agriculture garden is a kind of hints and metaphors of Hakka culture. The decoration and composition of these landscapes are the symbols and metaphors of Hakka culture by restructuring landscape elements.

2) Hakka characteristic red lantern

In the Hakka areas, red lantern is the continuation of the central plains culture, symbolizing family reunion, prosperity, happiness, brightness, vitality, perfection and wealth. In Yannanfei tea garden and Ketianxia Tourism Park, a large number of red lanterns not only highlight the Hakka characteristics and unique artistic style, but also supplement lighting in scenic spots. Red lanterns, as a common type of landscape symbols, strengthen the Hakka atmosphere in the scenic areas.

3) Concerted tradition of bamboo/wood materials

Hakka traditional architecture utilizes a large number of natural materials, such as bamboo, wood, red clay and stone, etc., while modern Hakka architecture uses much more artificial materials to meet the need of shaping and cost control. Nevertheless, these landscape architecture that is made of natural materials becomes the echo of modern Hakka landscape, nature and tradition. Thus, the reorganization of these materials can faintly reflect the silhouette of the traditional Hakka culture.

4) Regional specialty garden stone and stonework

Hakka traditional settlements are maily located in the hilly region which is more than 50% of Meizhou city. The special geological structure is teem with garden stone, granite and other kind of stone. The foundation of traditional Hakka building utilizes a large number of local stone. There are themic displays of garden stone in the three parks, such as the Yannanfei flying god rock in Yannanfei tea garden, King Stone in Ketianxia Tourism Park and many characteristic landscape stone among hiking trails in Ruishan mountain hi-tech eco-agricultural park. The local stone is also used at Hakka town in Ketianxia Tourism Park, the round square at entrance Stage one in Ruishan mountain hi-tech eco-agricultural park and lake landscape.

5) Healthy planting

A large number of healthy planting used at Ruixiang Valley in the Ruishan mountain hi-tech eco-agriculture park suit the theme of health and ecology. Under the background that health problems have gradually integrated into policies in many countries, more applications of healthy planting in landscape design are needed to meet the demand of the social development in China, especially in South China, which mathces the typical characteristics of Hakka culture that is eclectic, pioneering and enterprising, promoting the existence and development of the group and individuals of Hakka people. The first application of healthy planting in Hakka landscape is not only one of the attempts to apply the health problems in the landscape, but also landscape practitioners judgment of the development direction of the modern landscape under the influence of Hakka culture.

3 Conclusion

Hakka culture, especially its characteristics and spirit of being pioneering and enterprising, is a key factor in the continuation and longevity of the Hakka people's system. Under the guidance and influence of Hakka culture and its typical characteristics, combined with the existing typical landscape representation techniques, Hakka modern landscape shows a somewhat characteristics of the regional landscape. However, taking healthy landscape as the developmental goal and the main function will offer the Hakka modern landscape a more permanent life, which the author hopes that it can become regional characteristic landscape representative that guides one kind of modern landscape development direction.

References

1. Hagel, translated by Wang Zaoshi, Historical Philosophy, Shanghai Century Publishing Group, Shanghai Bookstore Press, 2001
2. Luo Jianhe, Cao Zhibo, "the inspiration of the form of traditional folk residence architecture and cultural spirit to modern logo design", "Guangdong education· Vocational Education", 2012
3. Xiong Jinhui, "the impact and influence of Hakka culture on contemporary design", "Journal of Jiangnan University", 2008
4. Luo Jianzi. "The inspiration of Hakka traditional architecture to modern architecture design – the design of Hakka world museum ", South China University of Technology Press, 1997
5. Wu Xichao finishing, " The abstract of thesis of Hakka spirit", the Chinese cities and counties network, 2008-07-31
   http://www.21ccct.com/sh/2008/0731/article_65.html
6. Cai Hong, "The perception of inheriting national culture and exploring Hakka architecture mystery -- on the new trend of architectural culture in Meizhou city"
7. http://www.kslaw.com.cn/paperview.asp?papert=article&id=4011， 2012-6-18
8. Li Junjie, "the protection and inheritance of the Hakka culture of world heritage landscape in Yongding Storied Buildings". Fujian Agriculture And Forestry University, 2013
9. Passengers City Network http://www.mz186.com
10. The website of Hi-tech industry garden at Meizhou, Guangdong http://www.gdmzgxq.gov.cn
11. Health News Network http://www.jkb.com.cn

AUTHOR INDEX

Abidin Shahriman Zainal....95
Aksas Rabia....16
Anwar Rusmadiah....95
Azrar Arab....16
Bae Jeong-hoon....55
Binti Idris Naila Rusma....36
Bt Mustaffa Zahiraniza....36
Challal Mouloud....16
Chen Fuqiang....70
Dehmas Mokrane....16
Gao Ruofei....110
Gijselhart Fons....82
Gunsing Jos....82
Hagemans Nyke....82
Hamed Yaman....36
Hassan Oskar Hasdinor....95
Hsu Ruei-Hung....88
Huang Hui....110
Huang Jun....1
Ismail Zuhaimy....30
Jeong Ji-Ho....45
Jiang Li-jun....104
Jonkers Hans....82
Kalynenko Mykyta....8
Kapteijns Bart....82
Kharoufi Hicham....45
Kim Soo-young....55
Kivits Eric....82
Klijn Peter....82
Kroeske Diederich....82
Langen Hans....82
Lee Jae-Chul....45
Li Zhe-lin....104
Lin Qichao....65
Liu Hu....1
Mahad Diyana Abdul....30
Oerlemans Bart....82
Oostindie Jan....82
Park Byeong-cheol....55
Qiu Han....70
Qiu Zhizhen....70
Sandfreni....22
Shafie A'fza....36
Shan Jianhua....70
Shin Sung-Chul....45, 55
Song Bo....104
Su Jing....104
Surendro Ir. Kridanto....22
Tian Yifeng....1
Trabelsi Mohamed....16
Tran Van-The....88
van Stuijvenberg Joost....82
Wang Huabin....65
Wang Huixing....70
Wang Xiaoyi....70
Wang Yuanyuan....76
Wu Yu-Ren....88